The Status and Distribution of New Jersey's Birds

The Status and Distribution of New Jersey's Birds

Charles F. Leck

Rutgers University Press
New Brunswick, New Jersey

Manufactured in the United States of America

Library of Congress Cataloging in Publication Data

Leck, Charles F., 1944–
The status and distribution of New Jersey's birds.
Bibliography: p.
1. Birds—New Jersey. I. Title.
QL684.N5L43 1984 598.29749 83–17655
ISBN 0–8135–1033–3

Contents

Acknowledgments

While unable to acknowledge all, I would like to list some of the people and organizations that have been principals of New Jersey ornithology, those whose field efforts and museum studies have been especially useful for an understanding of our avifauna. It is most difficult to be limited (my original list exceeded several hundred names!) so sincere acknowledgment is extended to the many unlisted birders as well as: C. C. Abbott, Vincent Abraitys, Horace Alexander, Harold Axtell, W. A. Babson, Peter Bacinski, James Baird, Roger Barton, John Benzinger, Mrs. Allen Black, Irving Black, Raymond Blicharz, Ed Bloor, Ned Boyajian, William Boyle, Alan Brady, John Bull, Joanna Burger, Cape May Bird Observatory, Edward Chalif, Frank M. Chapman, Ernest A. Choate, William S. Clark, Robert C. Conn, Bud Cooper, Allan D. Cruickshank, David Cutler, John Danzenbaker, Delaware Valley Ornithological Club, Howard Drinkwater, Peter Dunne, Warren F. Eaton, James Lee Edwards, Alfred E. Eynon, David Fables, Jr., Peter Fahey, Russell Fowler, Frank Frazier, Robert C. Frohling, Anne Galli, Ludlow Griscom, Fred Hamer, Gregory Hanisek, Leon A. Hausman, Donald S. Heintzelman, Joseph R. Jehl, Jr., Richard Kane, Tim Koebel, George Komorowski, Marie Kuhnen, Donald E. Kunkle, Dryden Kuser, Frank W. McLaughlin, Glenn Mahler, William Marx, James K. Meritt, May A. Middleton, Montclair Bird Club, Bertram G. Murray, Jr., New Jersey Audubon Society, Art Panzer, Joseph Penkala, Roger Tory Peterson, Julian K. Potter, Ted Proctor, Harold Rae, Charles H. Rogers, Richard C. Rosche, Richard Ryan, W. E. D. Scott, P. William Smith, Len Soucy, Tom Southerland, Edwin Stearns, Witmer Stone, Summit Nature Club, Clay Sutton, Jeff Swinebroad, Fred Tetlow, Stiles Thomas, Richard Thorsell, Fred Tilly, Trenton Naturalists Club, Charles Urner, Urner Ornithological Club, Wade Wander, David Ward, Watchung Nature Club, and Floyd Wolfarth. Many of these individuals and organizations have also been important in New Jersey conservation.

Within the above list three authors of regional avifaunas deserve particular praise: John Bull, David Fables, Jr., and Witmer Stone. Their works and their importance to the developing understanding of New Jersey's birdlife are respectfully acknowledged.

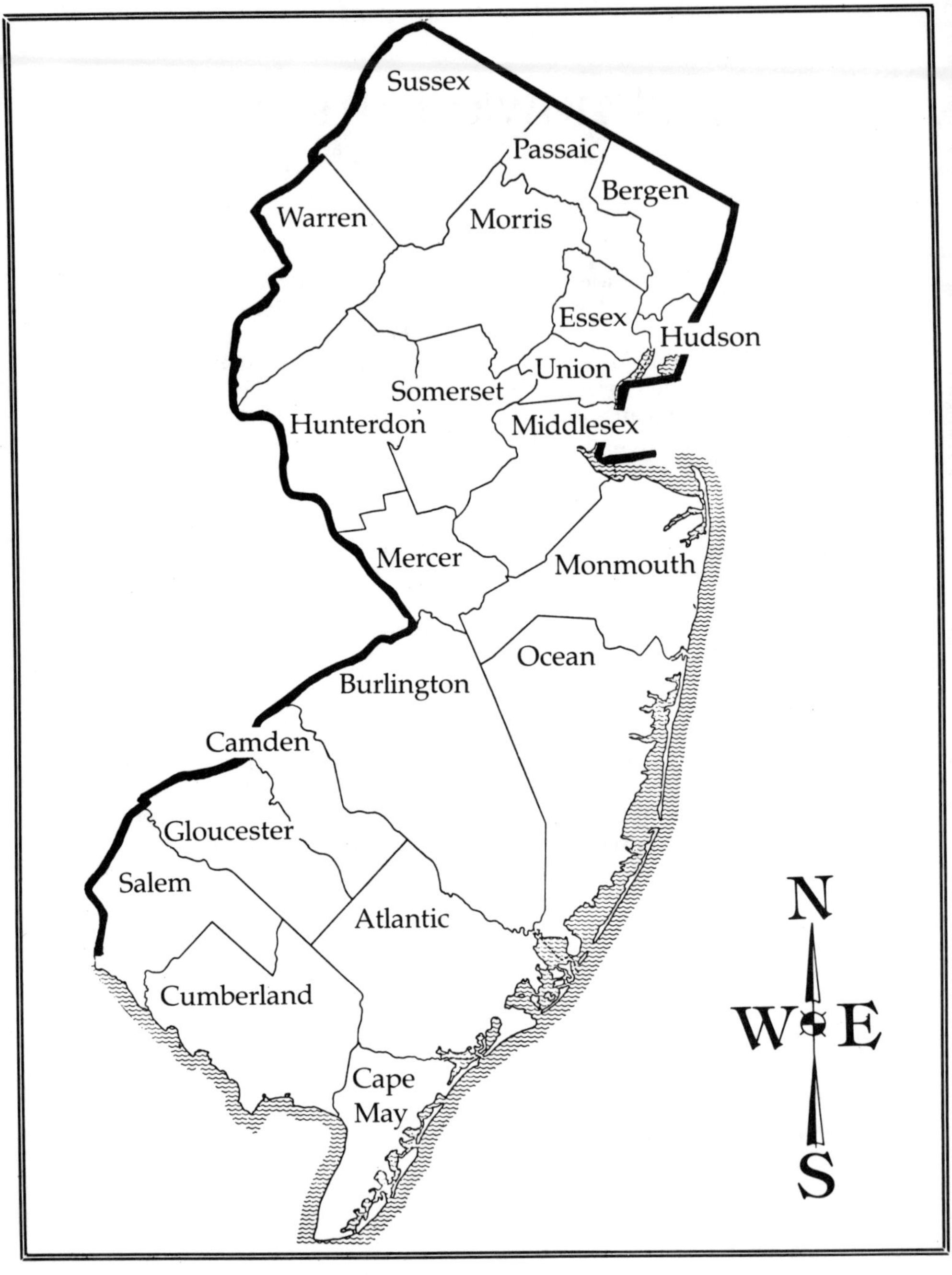
Sussex
Passaic
Bergen
Warren
Morris
Essex
Hudson
Union
Somerset
Hunterdon
Middlesex
Mercer
Monmouth
Ocean
Burlington
Camden
Gloucester
Salem
Atlantic
Cumberland
Cape May
N
W
E
S

Introduction

New Jersey has a remarkable diversity of birdlife. Well over 400 species have been recorded here. It is a diversity that is exceeded in perhaps a dozen much larger states, and only three have significantly greater species totals (Arizona, California, and Texas). The richness of New Jersey's avifauna reflects both a favorable geographical location and a wide variety of ecosystems. Our location, as a mid-latitude coastal state, provides for a notable mix of northern species with those of southern affinities (e.g. from Snowy Owls in winter to breeding Boat-tailed Grackles in summer). New Jersey's coast not only provides a wealth of marine birdlife, it also concentrates fall migration to spectacular levels (e.g. North America's greatest hawk migration occurs each autumn at Cape May). The relatively small size of the state belies its array of habitats, which are also important in the diversity of birdlife, especially for the breeding birds. From the Kittatinny Mountain ranges in the Northwest to the salt marshes of Cape May we find a series of natural environments that is surprisingly diverse (e.g. northern coniferous forest, bog, dry mountain ridge, swamps, lakes, rivers, old fields, hardwood forest, tidal freshwater marsh, southern coniferous forest, coastal vegetation, etc.). In addition to the birds already present in our natural ecosystems, many birds are attracted to managed or man-made habitats (e.g. reservoirs, agricultural fields, orchards, suburban yards, freshwater pools at wildlife refuges, parks, and even cities). Certainly few states could rival our mix of such a multitude of environments within such a limited area (approximately 7,500 square miles or 19,425 square kilometers).

Complementary to New Jersey's wealth of birdlife is a large human population that enjoys birds. Levels of interest vary greatly from individual to individual, but all are bonded by a shared appreciation of birds. The ways in which these avian interests are expressed also vary greatly: many people maintain bird feeders; others are nature photographers, bird-banders, ornithologists, hunters, traditional bird-watchers or birders, conservationists, and the like. It is hoped that the detailed information in the species accounts of this book will be of interest to all those who, in one way or another, are interested in New Jersey's birds. The bulk of the text provides a current summary of the status and distribution of all birds in our state. To many readers the accounts may simply provide enjoyable, informative reading. But there are also many who will *use* the accounts—for example, to compare their own local high counts with records elsewhere in New Jersey, or to determine the significance of their own rare bird sightings. It is hoped that in such ways this book will provide a substantial foundation for a bridge to our future knowledge and enjoyment of New Jersey's avifauna.

FORMAT OF THE SPECIES ACCOUNTS

The species names and most higher taxonomic units used throughout the text, as well as the order of presentation, follow the nomenclature used by the American Ornithologists' Union in the sixth edition of its checklist. For the higher groups, above the species level, I have adopted almost all the families recognized by the A.O.U. and have myself assigned common names to these families to aid most birders. In one case, however, I have not followed the new organization. The enormous Emberizidae family includes such a diverse assemblage of birds—warblers, tanagers, grosbeaks, buntings, sparrows, blackbirds, orioles, and crossbills—that to aid the reader I have followed the natural subunits of this array, using the subfamily for each grouping (e.g. Wood Warblers = *Parulinae*). Such groups were formerly given family status and will already be well familiar to bird students.

For the species names I have followed the A.O.U. for each element of the name: (1) the scientific name given in italics, (2) the person who first described the species, and (3) the common name. Where a second common name is in widespread use (e.g. in recently published field guides) I have often indicated that name as well. As a technical point it should be mentioned that the name of the describer, the original authority, is placed in parentheses when the original genus has been changed by subsequent taxonomic revision. It is particularly interesting to note how many of our New Jersey bird species were described by two notable persons—Linnaeus, the father of modern binomial nomenclature as used universally today, and Audubon, the father of American ornithology.

The species accounts begin with a range description, which is of course primarily concerned with the bird's status and distribution in New Jersey. Unidentified locations (e.g. "along the coast") are understood to involve reference to New Jersey as a state. Comments on the ranges outside of New Jersey have been given when such information helps in understanding the species status in our own state (e.g. if a species breeds as nearby as the Pocono Mountains of Pennsylvania). In general, for areas *outside* of New Jersey little distributional information is given. "United States" is taken to mean the biogeographic unit of the contiguous lower 48 states. (Where necessary, Alaska is referred to specifically.) And the term *North America* is generally used for the continental land mass of Canada and the United States, without reference to Greenland or Mexico. New Jersey localities, including towns, cities, parks, wildlife refuges, counties, bays, lakes, and mountains, are almost always well recognized features easily found on standard road maps (either commercial maps or the official state transportation map). Where reference is made to an obscure or little-known locality, or where there might be ambiguity, I have often given some additional geographic information, such as the county. Also, a gazetteer has been added for the lesser-known localities, and such locations are marked throughout the text with an asterisk. In many cases bird-watchers will already be familiar with even these otherwise obscure spots if they are of ornithological interest. (Detailed maps of many of the best birding areas of the state are available in the books listed later in this introduction, under "References for New Jersey Ornithology.")

Within the discussion of status I indicate the periods of peak abundance for migratory species (the vast majority of birds are migrants at some time of year). The periods of migration are given as general guides for the mass movement of each species. I have intentionally avoided the use of "earliest arrival dates" and "latest departure dates" as the extremes reflect aberrant birds, and such dates are continually being affected by more recent sightings of sick or stray birds. Rather, the *periods* of migration probably present an inclusive seasonal limit for the movement of 80–90 percent of all individuals for the species considered. In general there is often a marked difference in the seasonality of bird movements between our northernmost and southernmost counties—such differences are discussed, especially when they have a significant effect on distribution patterns.

Definitions of terms commonly used in the status discussions are in order. The term *resident* indicates a sedentary bird, and this term is usually modified by a seasonal adjective. For example, a summer resident is present in New Jersey during our summer at least (it may have various other seasonal statuses in the remainder of the year). A permanent resident is present year round, with little change in numbers other than breeding-season addition of young. (Relatively few species are permanent residents within our state.) In using terms of abundance I have tried to follow natural definitions, in general agreement with dictionary definitions and everyday use. Thus, in a declining series of abundance I would list: abundant, very common, common, uncommon, rare, and accidental. Occasionally other adjectives of abundance are used, but most of these are unambiguous (e.g. "very numerous" roughly equals "very common"). The term *casual* is often used interchangeably with *accidental* in reference to those rarest of birds that should not normally be occurring in New Jersey. In this series of abundance terms I have resisted the temptation to quantify each term although various authors have developed such schemes (e.g. "rare" could be defined as so many records per season). I have not adopted a quantified system of definitions for several reasons, all of which are biologically important. First, there can be no easy comparison of what would be common in terms of a numerical definition for such diverse groups as shorebirds, forest warblers, owls, ducks, and so on. Second, numbers of records increase with the increasing numbers of observers, and this could affect a bird's status in a quantified system, even if there were in fact no actual change in the bird's abundance. Third, even within the discussion of a given species it is helpful to have some flexibility in the use of a term for different seasons. (For example, Robins can certainly be considered common in summer, but they are also common in some winters at roosts along the coast—clearly the two "common" abundances could not be quantified as equal however.) Within a smaller geographical setting (e.g. a wildlife refuge) and especially within a single taxonomic group (e.g. waterfowl), a good case can be made for *some* comparative numbering systems, and comments from such work are included where they are considered helpful and reflective of the natural abundances.

For breeding species comments are provided on nesting periods and habitat, and often on nest construction. Where appropriate, the fact that a species has endangered status as a breeding bird within New Jersey is emphasized. I have particularly tried to indicate regional differences of breeding abundances (e.g. a species may be more common as a nesting bird in our northern counties). It should be noted, however, that breeding-bird distributions within our state are just now being studied in a systematic project. The current statewide breeding-bird survey should greatly contribute to our understanding of many species that have limited regional or local distributions. Ultimately, one may look forward to the publication of mapped distributions in a New Jersey breeding-bird atlas, as has been done in some other states.

Whenever possible I have included "high count" totals. These may represent high counts under any of a variety of conditions (e.g. migration flocks, winter roosts, breeding populations, etc.), but in each case the situation is indicated. I believe high count localities can often be of conservation interest for species that are either rare (e.g. an endangered songbird) or important in wildlife management (e.g. waterfowl). Thus I frequently give several high counts, representing several areas. A temporal series of high counts is also useful to show major population trends within a species. Alas, for many birds no high count data are available for New Jersey—so do make counts!

Certain standard bird censuses are referred to in the text. These include the annual Christmas counts conducted about the end of December each year through the National Audubon Society, breeding-bird censuses by various groups, and National Wildlife Refuge counts made by the Federal Fish and Wildlife Service. Our state has a strong tradition of conducting many such censuses over a series of many years, a tradition that provides us with a historical perspective of avian population changes.

In addition to its natural resources New Jersey can boast a long tradition of naturalists, and the birders are legion. The records that went into this volume include field sightings from over a century made by more than 1,000 observers. (As suggested earlier in the introduction, there are many thousands of bird-watchers of varying levels of activity in the state today.) I have adopted a modern system of presenting records without observers' names for several reasons. Foremost is the space saving: publishing them has become prohibitively expensive. The editor of *American Birds,* our leading magazine of field reports, has made an eloquent argument for publishing sightings without observers' names, and this will certainly be a future trend, at least in statewide books. (Those wishing to determine the source of any report within the text can usually refer to the original—in almost all cases *New Jersey Nature News* or its successor *New Jersey Audubon: Records of New Jersey Birds.*)

Under some species accounts, usually at the conclusion, there are literature citations to research articles on those species. Such citations are given as the author's last name and the date of publication (e.g. Smith 1983). Full information on the reference, including the title of the article and the original journal or magazine, is given in the bibliography.

ANALYSIS OF THE AVIFAUNA

The avifauna of any area as large as a state is dynamic—there are always increases and decreases, additions and losses. The main text of species accounts includes 450 species, with 414 of these well confirmed in New Jersey, as with a specimen or multiple observations by many observers. The basic 414 species include a small number of birds that are no longer part of our avifauna because of extinction (e.g. Eskimo Curlew), the long-term failure of introduced populations (e.g. Skylark), or other natural declines. The continual growth of the list of bird species recorded in New Jersey is reflected in the published totals of Stone (1909) with 340 species, Fables (1955) with 385 species, and the present list with a very conservative 414 (up to at least 450 species if all records are accepted). The continual addition of species is made possible by a variety of factors: (1) new species become available through range expansions and introductions, (2) bird-watching efforts have expanded greatly, as with the Cape May hawk watch, (3) new locations are being explored by birders (e.g. pelagic trips). In addition, the identification skills of birders are improving, and they are alert to such subtle possibilities as the Sharp-tailed Sandpiper.

The difference between the 450 and 414 species mentioned above reflects 36 species for which we need better confirmation of their occurrence to consider them fully documented for New Jersey. These 36 species are listed below, and there is little doubt that almost all have actually strayed to our state, in the past if not in the present. Waterfowl present a particular problem as there can be escapes from regional collections, leaving the status of an exotic duck in the wild as questionable. There *may* be errors of identification in the reports of a few of these unconfirmed species, but most of these species will occur again in New Jersey in the future—then we should be careful to get good documentation. The main text discusses each of these.

The 36 species are: Arctic Loon, Yellow-billed Loon, Yellow-nosed Albatross, Black-browed Albatross, Black-capped Petrel, British Storm-Petrel, Band-rumped Storm-Petrel, Olivaceous Cormorant, White-faced Ibis, Lesser White-fronted Goose, Barnacle Goose, Masked Duck, European Sparrowhawk, Whooping Crane, Greater Golden-Plover, American Black Oystercatcher, Spotted Redshank, Rufous-necked Stint, Sharp-tailed Sandpiper, Great Snipe, Mew Gull, California Gull, White-winged Tern, Great Auk (extinct), Band-tailed Pigeon, White-winged Dove, Carolina Parakeet (extinct), Smooth-billed Ani, Great Gray Owl, Western Flycatcher, Brown-crested Flycatcher, Gray Jay, Sage Thrasher, Sprague's Pipit, Cassin's Sparrow, and McCown's Longspur.

There will, of course, always be some species for which we lack critical confirmation, but these are rarities that play only a very small role in the overall drama of our state's avifauna. We can even predict some future species to be recorded by New Jersey birders—in the early fall of 1982 I listed as five probable accidentals that might soon be added to the state list: Rufous Hummingbird, Mountain Bluebird, Kirtland's Warbler, Baird's Sparrow, and Smith's Longspur. (The Mountain Bluebird *was* observed and photographed at Brigantine Refuge* later that fall.) More significant, however, are the occurrences of the first few invaders of increasing species that will subsequently become an important part of our avifauna. Such early sightings allow us to trace the history of expansion in such birds as the Cattle Egret and House Finch, and this provides us with a better understanding of the population dynamics of invasions. Over the last two decades New Jersey has averaged almost two additions per year to the total list of bird species recorded in the state, but such a rate of increase cannot be sustained into the next century. (There are obvious natural limits to the potential pool of new species available—and modern birders, with great enthusiasm, are pressing toward those limits.) Recent additions are thus all quite exotic: for example, White-tailed Tropicbird, Chestnut-collared Longspur, Townsend's Solitaire, Black-capped Petrel, Ferruginous Hawk, and European Sparrowhawk (See Table 1).

If the rarities form the "extras" of our state's avifauna, the breeding birds form the backbone. This element includes at least

TABLE 1

Additions to New Jersey Avifauna, 1960–1981

1960	Great Kiskadee
1961	Fulvous Whistling-Duck, Cassin's Sparrow, Black-throated Sparrow
1962	Rufous-necked Stint (?), Boreal Owl, Virginia's Warbler, Golden-crowned Sparrow
1963	Great Snipe (?)
1965	Spotted Redshank
1966	Tufted Duck
1967	California Gull
1968	Monk Parakeet
1969	Great Skua†
1971	Anhinga, Lesser White-fronted Goose (?), Black-tailed Godwit, Townsend's Warbler
1972	Black-browed Albatross, Ross' Goose, Eurasian Kestrel
1975	Manx Shearwater, Sharp-tailed Sandpiper (?), Franklin's Gull
1976	MacGillivray's Warbler†
1977	Yellow-billed Loon (?), White-faced Ibis, South Polar Skua†
1978	European Sparrowhawk (?), Ferruginous Hawk
1979	Black-capped Petrel
1980	Band-tailed Pigeon (?), Townsend's Solitaire, Chestnut-collared Longspur
1981	White-tailed Tropicbird, White-winged Dove (?)

(?) Indicates species for which there is some question about origin (wild or escaped?) or some problem of identification.

† Indicates species that may have been recorded earlier than the year given.

198 species that are documented as current or recent breeding birds within New Jersey. Just over half (105) of these are passerines or "songbirds," and most are present in the state only during the breeding-season months of late spring and summer. In spite of a variety of environmental losses, our state has had a rather continual increase in the number of nesting species, and only a few breeding species have been lost (e.g. Wilson's Plover). The additions to our list of breeding species are presented in Table 2. The list shows a period of addition of duck species and other waterbirds (1941–1960), followed by another set of two decades during which northern landbirds and reintroduced species are important. (The northern species are especially associated with pine plantations in New Jersey.) There are certainly a few species expected as future additions to the nesting avifauna; possible species include: Double-crested Cor-

TABLE 2

Additions to the New Jersey List of Breeding Species since 1920

1920	Mute Swan (introduction)†
1922	Chuck-will's-widow
1941	Red Crossbill
1946	Herring Gull
1949	Gadwall, Hooded Merganser
1950	Northern Shoveler
1955	Glossy Ibis
1957	Blue-winged Teal
1958	Cattle Egret, Ruddy Duck, Gull-billed Tern (first breeding in modern times)
1959	Canada Goose†
1960	Green-winged Teal, House Finch (introduced)†
1962	Pine Siskin, Evening Grosbeak (only one year)
1963	Golden-crowned Kinglet
1966	Great Black-backed Gull
1968	Monk Parakeet (introduced)
1971	White-throated Sparrow
1972	Northern Goshawk†
1973	Dickcissel (first breeding in modern times)
1977	Wild Turkey (reintroduction)
1978	Common Merganser†
1979	Yellow-rumped Warbler
1980	Peregrine Falcon (reintroduction)
1981	Black Vulture

† Indicates species that may have bred earlier than the year given.

morant, White Ibis, American Wigeon, Mississippi Kite, Purple Gallinule, Ring-billed Gull, Royal Tern, Northern Saw-whet Owl, Yellow-bellied Sapsucker, Olive-sided Flycatcher, Yellow-bellied Flycatcher, Ruby-crowned Kinglet, Blackpoll Warbler, Mourning Warbler, and Clay-colored Sparrow. (A few of these, such as the Cormorant, may already have bred in the state since 1980.) As a conservation note it should be emphasized that New Jersey has low populations of a wide variety of species from our list of breeding birds. Such species with declining or tenuously held populations deserve careful monitoring and protection. Included in this group of endangered breeders are some of the birds of prey and some grassland species, for example. (These problems are discussed within the main species accounts.)

Readers wishing to look further into New Jersey birdlife can consult books that have reviewed the state's avifauna, or at least parts of it, at various times since 1909 (the date of the first book on New Jersey birds). These books, some of which are listed below, provide the foundation of our knowledge of historical changes in the avifauna and remain valuable references long after they were written. (Full bibliographical details can be found in the bibliography.)

Bull, John. 1964. *Birds of the New York area.* Includes northern half of New Jersey.

Fables, David G., Jr. 1955. *Annotated list of New Jersey birds.*

Leck, Charles. 1975b. *The birds of New Jersey: Their habits and habitats.*

Stone, Witmer. 1909. *The birds of New Jersey.*

———. [1937] 1965. *Bird studies of Old Cape May.*

In addition, two modern reviews of our neighboring New York and Pennsylvania avifaunas are quite useful for study and comparison with our New Jersey work. These are: John Bull, *Birds of New York State* (1974, supplement 1976), and Merrill Wood, *Birds of Pennsylvania* (1979).

The following list contains books that can help bird-watchers find favorable areas for field observations throughout the state. (Several deal with other states besides New Jersey.)

Akers, James F. 1981. *All year birding in southern New Jersey.*

Arbib, R. S., O. S. Pettingill, and S. H. Spofford. 1966. *Enjoying birds around New York City.*

Boyle, William J., Jr. 1979. *New Jersey field trip guide.*

Harding, John J., and Justin J. Harding. 1980. *Birding the Delaware Valley region.* An excellent guide for areas of western and southern New Jersey as well as adjacent states.

Heintzelman, Donald S. 1972c. *A guide to northeastern hawk watching.*

Pettingill, Olin S. 1977. *A guide to bird finding east of the Mississippi.* The latest edition of this classic includes a good section on New Jersey.

In addition to the above works there are many other types of ornithological publications too numerous to mention, from local checklists to booklets. Some of these, like Donald S. Heintzelman's *Hawks of New Jersey* (1970) and Leonard J. Soucy, Jr.'s *New Jersey's owls* (1980), are primarily of interest to the specialist. Ornithological sections of regional natural history books can be quite informative, and nonornithological state publications can help one locate favorable habitats for birding. One of the best guides is a recently revised handbook that is much used by fishermen and hunters: Steve Perrone, ed., *Guide to wildlife management areas* (1982).

For those interested in detailed field studies of our New Jersey birds a bibliography is provided at the end of this book which lists, in addition to the works mentioned in this section, articles from the recent ornithological literature. It is hoped that many birders will enjoy the additional pleasure of making field studies, and such journal articles provide an introduction to the methods and goals of modern ornithological research. Details of the natural history of our birds are, indeed, fascinating.

Species Accounts

LOONS / *Gaviidae*

RED-THROATED LOON *Gavia stellata* (Pontoppidan)

The Red-throated Loon is a common migrant and winter resident, along the coast. Migration peaks from late March to mid-April and from mid-November to early December. During migration this species usually outnumbers the Common Loon along the shore.

High counts include 1,500 in one hour migrating southward at Stone Harbor on November 17, 1978, about 2,000 at Cape May Point on April 15, 1963, and "most" of 7,000 loons counted at Avalon on November 18, 1979. Inland it is relatively scarce; a good count was obtained at Round Valley Reservoir* with 35 birds during mid-October 1972.

ARCTIC LOON *Gavia arctica* (Linnaeus)

This Pacific species is very rarely reported from New Jersey in winter, usually along the coast. Unfortunately the field identification of the winter-plumaged bird can be quite difficult (e.g. confusion with the small races of the Common Loon). In most years there are no records; it would be most desirable to obtain good field photographs of any loon suspected to be this species.

Reports include the Atlantic Highlands waterfront, Monmouth County (February 5, 1938), Manasquan Inlet* (February 11, 1952), Shark River Inlet* (April 14, 1956), and inland at the Round Valley Reservoir* (November 16–December 3, 1978).

COMMON LOON *Gavia immer* (Brünnich)

This loon is common as a migrant over most of the state, and it will pause at larger lakes and reservoirs. It is also a winter resident along the Atlantic coast and the Delaware Bay shore. Migration peaks from April to early May and from mid-October to mid-November. (Migrants are often seen inland flying overhead.) A few summer nonbreeding birds are sometimes found in northern New Jersey; the nearest breeding is now in the Adirondacks of New York, although it formerly nested as close as the Poconos of Pennsylvania (until 1955, and again in 1982).

High counts include 70 at Swartswood Lake* (April 28, 1953), and 150 flying by Raccoon Ridge* (April 16, 1976). A seasonal high tally of 628 was recorded at Raccoon Ridge* during the spring migration of 1976.

YELLOW-BILLED LOON *Gavia adamsii* (Gray)

This arctic species has only been recorded accidentally in New Jersey, with a single sighting at Shark River Inlet* on December 8, 1977. Subsequent searching failed to find the bird, and until a confirmed record is obtained the species is listed as hypothetical. (There is a Long Island, New York, specimen from 1930.) A very detailed report of the New Jersey sighting was eventually published (*Urner Field Observer,* 1983, 18: 35–39).

GREBES / *Podicipedidae*

PIED-BILLED GREBE *Podilymbus podiceps* (Linnaeus)

Today the Pied-billed Grebe is a very uncommon breeding bird in freshwater ponds and marshes, mostly in the northern half of the state. Formerly it was a common summer resident in many areas of New Jersey. Migration peaks from mid-March to mid-April and from mid-September to mid-November. A few remain into winter, particularly near the coast. Nesting occurs in May and June, but

New Jersey has had declining populations through the 1970s. A recent high concentration of breeding birds was at Kearny Marsh* in 1979, with more than a dozen pair. In 1981 there were only two known breeding sites in New Jersey: at Kearny Marsh* (where it is relatively common), and at Pedricktown (only one pair). The freshwater nesting habitats are limited as the grebes prefer bodies of water that have considerable aquatic vegetation and adjacent expanses of open water. Throughout much of its range in North America it is thus only a scattered or local breeding species.

HORNED GREBE *Podiceps auritus* (Linnaeus)

The Horned Grebe is a coastal species, common as a transient and winter resident. Migration peaks occur from mid-March to mid-April and during the first half of November. Away from the coast it is seen less commonly, usually on larger lakes or reservoirs. A few nonbreeding birds are sometimes present as summer stragglers.

A count of 80 at Round Valley Reservoir* during mid-October 1975 was especially impressive for an inland locality.

RED-NECKED GREBE *Podiceps grisegena* (Boddaert)

This grebe is usually rare as a transient and winter visitor, along both the coast and the Delaware River. Infrequently there are invasion years (e.g. 1958–1959) when it is more common, with reports of up to a maximum of 90 a day in the state. In the invasion of early March 1958 there were several Red-necked Grebes under almost every bridge along the Delaware River. Single locality high counts include 10 at Barnegat Light (December 3, 1963), and 17 at Swartswood Lake* (April 5, 1959).

EARED GREBE *Podiceps nigricollis* Brehm

This grebe is primarily a bird of the western states, and it occurs only rarely in New Jersey. Usually there are no more than one or a few records within the state each winter, from December to March. Recent sightings include an April 1965 record from Brigantine National Wildlife Refuge,* and Cape May reports (December 29, 1968, and more recent Christmas counts).

WESTERN GREBE

Aechmophorus occidentalis (Lawrence)

As suggested by its name, this grebe is also from the western states. It is a very rare visitor to New Jersey's coast with only one or two records every several years, from January to March. (It is thus decidedly rarer than the Eared Grebe.) The locality with the most reports is Barnegat Inlet* (e.g. December 28, 1958, and February–March 1970).

ALBATROSSES / *Diomedeidae*

YELLOW-NOSED ALBATROSS

Diomedea chlororhynchos Gmelin

Hypothetical? There are no confirmed reports of this pelagic species for New Jersey waters. However it has occurred off Long Island, New York, and it should be looked for on far offshore fishing trips that go to the edge of the continental shelf in the area of the major submarine canyons. New Jersey's first albatross sighting came in the summer of 1967; an "albatross species" was seen off Cape May on June 4. Several birds were reported as albatross species at the Hudson Canyon,* from captains at this far offshore area in June 1972. (This species and the next travel great distances in approaching our coast; they both breed on remote islands in cold-water areas of the southern hemisphere.)

BLACK-BROWED ALBATROSS

Diomedea melanophris Temminck

This albatross is an accidental or possibly very rare visitor to our far offshore waters in summer. Its status and distribution in this region are uncertain because we lack sufficient pelagic fieldwork. There have been observations of single birds from the Hudson Canyon* (May and early June during the 1970s). A maximum of 2 (an adult and an immature) was recorded *from shore* at Cape May

Point on October 7, 1974. (In 1972 there were records of 2 birds each from Buzzards Bay, Massachusetts, and offshore North Carolina.)

SHEARWATERS AND PETRELS *Procellariidae*

NORTHERN FULMAR *Fulmaris glacialis* (Linnaeus)

This North Atlantic seabird visits our offshore waters, primarily in the spring (April–May) and fall (October–November). Our birds are usually of the light phase, with whitish head and underparts. The species has had expanding populations in this century, in both North America and Europe. With this increase New Jersey is having more frequent sightings, even during the summer months. High counts come from offshore fishing fleets (150 were recorded at the scallop grounds 60 miles east of Island Beach* on May 26, 1978).

BLACK-CAPPED PETREL *Pterodroma hasitata* (Kuhl)

This petrel breeds in the Greater Antilles and is generally restricted to warm ocean waters. Its New Jersey status is uncertain, or hypothetical, until better information is available. The report nearest to our state was made from the far offshore zone, 135 miles southeast of Ship Bottom (closest land), on July 18, 1979. This pelagic species might be a regular albeit very rare visitor in currents of the Gulf Stream during late summer and fall. It is, for example, regularly reported from deep waters off North Carolina. Currently efforts are being made to find this species far to the southeast of Cape May (e.g. pelagic trips to the Baltimore Canyon*).

CORY'S SHEARWATER *Calonectris diomedea* (Scopoli)

This offshore summer visitor occurs with about equal frequency during the months of May through October, except for modest migration peaks (early June and late October). High counts come from far offshore and include 78 about 50 miles east of Atlantic

City (June 21, 1979) and 200 at the Hudson Canyon* (June 8, 1974). Birders on fishing boats near the coastline, even within sight of land, sometimes report small numbers through the summer. (The Cory's Shearwater is particularly associated with groups of porpoises.)

GREATER SHEARWATER *Puffinus gravis* (O'Reilly)

This offshore visitor is found irregularly between mid-May and mid-November. The spring migration is much more pronounced than the fall flight as there is a general movement to the eastern Atlantic in late summer. (Breeding is restricted to a few South Atlantic islands.) The May–June migration peak can include large flights with thousands passing New Jersey over deep waters. This species is often seen by offshore fishermen as it comes among the boats for fish scraps.

High counts include 850 at a squid fleet 70 miles off Cape May during mid-July 1979, 1,250 at scallop fishing grounds 70 miles east of Corson's Inlet* (May 27, 1977), and 2,500 on a pelagic trip from Barnegat Light to the Hudson Canyon* (May 30, 1981).

SOOTY SHEARWATER *Puffinus griseus* (Gmelin)

This pelagic visitor is usually seen offshore, but sightings from land are also regularly reported. The chances of seeing this and other such ocean birds from shore are greatly enhanced by strong easterly or northeasterly winds that blow the shearwaters toward the coast. (Still, it is rare to see more than a few anywhere except on the high seas.) This shearwater is primarily a migrant in our region with peak numbers during the period of mid-May to mid-June. There is a smaller movement in late August, and a few birds are reported in other months (July, and September into October). A maximum of 2,500 was seen at scallop fishing grounds, 70 miles east of Corson's Inlet* on May 27, 1977. This species, like the Greater Shearwater, is particularly attracted to fish scraps and pieces of bait from fishing boats—indeed the birds can be a nuisance in fouling fishing lines!

MANX SHEARWATER *Puffinus puffinus* (Brünnich)

For New Jersey this shearwater is primarily a very rare pelagic visitor from European waters. Most sightings occur in May, but records from nearby states include the months from April to September, and even December. The first breeding record for this species in North America came from coastal Massachusetts in 1973, and by 1980 it was apparently nesting on numerous islets from Rhode Island northward. The first New Jersey report came from the Hudson Canyon* on May 31, 1975 (a maximum of 3 was established here on May 29, 1976). Rare reports have come from recent pelagic trips to this submarine canyon (e.g. May 30, 1981), and even from the Cape May Ferry* (June 17, 1981).

AUDUBON'S SHEARWATER

Puffinus lherminieri Lesson

This ocean visitor from tropical seas is usually quite rare except with offshore invasions of warm Gulf Stream currents (i.e. *Sargassum* seaweed waters). Almost all of our records are from deep-water areas between mid-August and mid-October. High counts include 173 on July 18, 1979, from 130 miles offshore, where it was the most common bird in the Gulf Stream. Another remarkable count was made on September 19, 1976, with 225 seen about 100 miles southeast of Barnegat Light.

STORM-PETRELS / *Hydrobatidae*

WILSON'S STORM-PETREL *Oceanites oceanicus* (Kuhl)

The Wilson's Storm-Petrel is often considered one of the world's most abundant birds, and although it breeds far off in the South Atlantic and Antarctic it moves northward to become a common ocean bird off New Jersey's coast from mid-April to early October each year. Peak numbers are usually reached in August, although there can be considerable variations in apparent abundance from year to year, depending on regional movements of ocean currents. The species is easily attracted (by olfaction!) to chum or chopped fish thrown from fishing boats. Since the 1970s, however, many captains are relying more on electronic fish finders than chumming, and without such baiting the petrel can be easily missed on a boat trip. High counts include 40,000 at the Hudson Canyon* (May 28, 1983), 3,000 a few miles off Sea Girt (July 7, 1973), more than 4,000 on a Hudson Canyon* pelagic trip (May 30, 1981), and 5,000 in the lower New York Bay off Monmouth Beach (July 19, 1973).

WHITE-FACED STORM-PETREL

Pelagodroma marina (Latham)

This seabird of the South Atlantic occurs only as an accidental offshore visitor to New Jersey. Most records come during the late summer period of migration, in August and September. Sightings include one 9 miles off Avalon (September 5, 1973), one 90 miles southeast of Barnegat Light (September 19, 1979), and two 90 miles southeast of Cape May (September 10, 1980).

BRITISH STORM-PETREL

Hydrobates pelagicus (Linnaeus)

This European species is recorded only as an accidental visitor off the North American coast, primarily in the Northeast (e.g. Maine and Nova Scotia). A possible New Jersey record comes from the Hudson Canyon* on September 17, 1972, where a small dark storm-petrel was tentatively identified as this species. It will retain a hypothetical status for our state until a good photograph or specimen is obtained with future ocean research.

LEACH'S STORM-PETREL

Oceanodroma leucorhoa (Vieillot)

Although this species nests as nearby as islands along the Maine coast it is uncommon or rare as a pelagic visitor off New Jersey. Our records generally come from May to October, and this species seems to be more restricted to deep waters than the more common Wilson's Storm-Petrel. Very rarely the Leach's Storm-Petrel becomes land blown in strong storms or severe gales, as in August 1842 when there were a number of inland reports. A high count of 17 was made far offshore at the Hudson Canyon* on May 27, 1973.

BAND-RUMPED or HARCOURT'S STORM-PETREL

Oceanodroma castro (Harcourt)

The New Jersey status of this tropical seabird is uncertain, but it is probably a casual pelagic visitor in summer. One state (?) record far at sea came from bird-watchers on a boat 185 miles southeast of Barnegat Inlet.* This bird was within a group of Leach's Storm-Petrels so comparison with its most similar species was possible. Even so, field identification of the Band-rumped Storm-Petrel is difficult at best because of the close similarities in appearance of the various storm-petrel species (not to mention the effects of boat movements on binocular viewing).

TROPICBIRDS / *Phaethontidae*

WHITE-TAILED TROPICBIRD

Phaethon lepturus Daudin

A seabird of tropical oceans, the White-tailed Tropicbird breeds no closer to our coast than Bermuda. There is only a single sight record for New Jersey by four observers at North Wildwood (Great Channel*) on October 8, 1981. There are also some records from nearby states.

RED-BILLED TROPICBIRD

Phaethon aethereus Linnaeus

This seabird of tropical oceans is strictly accidental on the Atlantic coast of North America. New Jersey's first and only record is of an immature bird that washed ashore (in oiled condition) at Seaside Park on May 24, 1983. (There is a June 1963 record from Long Island, New York.)

GANNETS AND BOOBIES / *Sulidae*

BROWN BOOBY *Sula leucogaster* (Boddaert)

This tropical seabird is an accidental visitor in our offshore zone, usually in late summer or early fall. Caution in field identification is required as this species could be confused with the immature Gannet. The Brown Booby should be searched for after hurricanes, as noted on Long Island, New York (1936, 1946). Recent New Jersey reports include 2 adults off Island Beach* (October 22, 1979), 3 off Manasquan Inlet* (June 3, 1973), and an immature at Cape May Point (May 26, 1980).

NORTHERN GANNET *Sula bassanus* (Linnaeus)

After breeding in the maritime provinces of Atlantic Canada the Gannet migrates south to winter along the United States coast. New Jersey has migration peaks of this ocean transient in March–April and November. Lesser numbers are seen as winter visitors. (At most times they are barely within sight of land, and telescopes greatly improve observations.) As is the case with other pelagics, easterly winds and foul ocean weather can drive this species closer to shore. During migration there can also be some movement into Delaware Bay.

High counts include 200 at Point Pleasant, Ocean County (November 12, 1927), and 1,458 migrating past Avalon (November 23, 1979).

PELICANS / *Pelecanidae*

AMERICAN WHITE PELICAN
Pelecanus erythrorhynchos Gmelin

Although the White Pelican breeds in western North America its migration route takes it to the Southeast (Texas to Florida), and strays occasionally occur along the upper Atlantic coast including New Jersey's shore. All the New Jersey records are from coastal areas, the first state sighting coming from Sandy Hook in 1864. Subsequent reports include Beverly (September 25, 1943), Barnegat Bay* (November 10, 1946), Beach Haven Inlet* (August 18–30, 1949), Stone Harbor (November 7–14, 1951), Normandy Beach (October 13 and 20, 1957), and Brigantine National Wildlife Refuge* (September 6–December 1978 and again in April 1979). The last bird was particularly evident and was enjoyed by hundreds of observers. Recently, one was at the Great Bay* (May 30, 1982).

BROWN PELICAN *Pelecanus occidentalis* Linnaeus

Although the Brown Pelican nests as nearby as coastal North Carolina, it is only a very rare visitor to New Jersey. Most reports are from our southeastern coast. The first state record goes back to 1837, from Sandy Hook. During the first half of this century there were 7 New Jersey sightings, 3 from May and the remainder from autumn (August–November). Recent records include Normandy Beach (August 5, 1969), an immature at Cape May Point in late January 1977, offshore sightings at Sandy Hook and Asbury Park (May 15–16, 1981), 3 from a pelagic trip 42 miles southeast of Cape May (October 4, 1981), a group of 13 at Cape May (June 7, 1982), and an unprecedented invasion of more than 100 along the Delaware Bay shore in the summer of 1983.

CORMORANTS / *Phalacrocoracidae*

GREAT CORMORANT *Phalacrocorax carbo* (Linnaeus)

Formerly this northern seabird was only a rare winter visitor to our shore. Today it is regular in small numbers from October to May because it has had a southward expansion of its winter range. The Great Cormorant normally frequents rocky jetties at coastal inlets, but there are a few inland records from large lakes (e.g. Lake Assunpink*).

The first New Jersey record was an 1883 specimen from Bridgeton. Positive records for southern New Jersey came in 1929 and

1931, and it had become regular along the coast by 1955. High counts include 8 at Cape May (April 2, 1980), 10 at Barnegat Light (February 17, 1961), and 25 migrating past Sandy Hook (October 25, 1981).

DOUBLE-CRESTED CORMORANT

Phalacrocorax auritus (Lesson)

A common coastal migrant, the Double-crested Cormorant passes in peak numbers during April and September–October. Small numbers remain into winter or summer, usually as individuals. It is also infrequent inland on lakes or rivers, during migration. This species nests near our state to both the north and south, and in recent years the behavior of summering birds suggests New Jersey breeding, at least in the near future. (A bird carrying nesting material was seen in Barnegat Bay on July 2, 1981.) High counts include 700 on the Delaware River (August 25, 1977), 3,000 past Island Beach* (October 20, 1967), 7,740 along Avalon (October 15, 1979), and about 14,000 at Cape May (October 11–12, 1981).

OLIVACEOUS CORMORANT

Phalacrocorax olivaceus (Humboldt)

Hypothetical: a report by two independent observers at Cape May Point on October 3, 1980, has been considered most tentative. This tropical species normally ranges northward only to the Louisiana coast. Identification problems could be caused by small individuals of the Double-crested Cormorant.

ANHINGAS / *Anhingidae*

ANHINGA

Anhinga anhinga (Linnaeus)

This tropical species breeds north only to North Carolina, and is strictly an accidental visitor to our region. The single New Jersey record is of a soaring bird at Cape May on September 25, 1971. (It was seen by many observers and was well documented.)

FRIGATEBIRDS / *Fregatidae*

MAGNIFICENT FRIGATEBIRD *Fregata magnificens* Mathews

The frigatebird is generally restricted to tropical seas, but as one of the world's best fliers it occasionally wanders northward. It is a very rare visitor to the New Jersey coast, and older records often reflect birds brought here by hurricanes. State sightings include an 1877 record, 2 Cape May birds (August 3, 1926, and September 7, 1981), and 1 each at Brigantine (September 15, 1936), Avalon (June 8, 1949), Shark River* (June 20–21, 1969), Belmar (November 4, 1976), and Northvale in Bergen County (April 7, 1982).

BITTERNS AND HERONS / *Ardeidae*

AMERICAN BITTERN *Botaurus lentiginosus* (Rackett)

The American Bittern is an uncommon permanent resident, with fewer remaining in the winter months. It nests quite locally, usually in our larger freshwater marshes, from May to June. This shy species is seen most frequently during migration, when it may become relatively common in marshes and lakeshore vegetation throughout the state.

A high count of 50 was reported from an estimate of calling birds at the Great Swamp* in the summer of 1973.

LEAST BITTERN *Ixobrychus exilis* (Gmelin)

This tiny heron is an uncommon summer resident from May to September. As a local breeder it is limited to large freshwater or brackish marshes (e.g. Manahawkin, Trenton Marshes,* Kearny Marsh*), where it nests from May into July. The species is not doing well, and it has been lost from many former breeding localities (e.g. Princeton). (The solitary nests are often placed in cattails.)

As with the larger bittern, the high count for this species comes from the Great Swamp* with an estimate of 50 in the summer of 1973. I am unaware of any recent New Jersey sightings of the rare chestnut phase of this species, known as the "Cory's" Least Bittern.

GREAT BLUE HERON *Ardea herodias* Linnaeus

This large heron is generally an uncommon permanent resident, most numerous during periods of migration (March–April and September–October). Winter birds are usually limited to the coastal areas, and even there they are scarce. As in other areas of its breeding range locally the Great Blue Heron nests in small colonies from March to June, and many colonies have declined or been lost in recent decades. None was reported in the 1979 survey of coastal New Jersey's breeding colonies. There are four small rookeries in wooded swamps at inland localities (the state's breeding population was estimated at 291 pair in 1978). The Great Swamp* had a new colony start in 1982.

A high count from Brigantine National Wildlife Refuge* was 39 on October 12, 1979. There is an unsubstantiated report of the white form, the "Great White" Heron, from the Brigantine Refuge.*

GREAT or COMMON EGRET *Casmerodius albus* (Linnaeus)

Our largest egret is primarily a coastal resident in summer (March through November). A few individuals remain in winter in southern areas of the state. Nesting occurs from April to June, mostly on uninhabited islets of coastal bays. (The bulky stick nests are placed fairly high in the vegetation, even in sizable trees.) In this century breeding started in New Jersey in 1928 (the species had been eliminated as a breeding bird in the nineteenth century by plume hunters). Its nesting spread as far north as Sandy Hook by 1955. Today we see encouraging high counts such as 450 at Stone Harbor (summer 1966), and an impressive 598 on a coastal New Jersey survey (summer of 1979).

SNOWY EGRET *Egretta thula* (Molina)

The Snowy Egret is primarily a coastal resident in summer (April through October), with small numbers remaining into winter. (It is the egret most commonly reported on New Jersey Christmas counts; the state total can be several hundred.) Most nesting occurs on salt-marsh islands that have shrubbery and trees. Its population has a history similar to that of the Great Egret, with a notable decline at the end of the nineteenth century. Records of old colonies with Snowy Egrets date from around 1872 to 1888, with only a single pair in the last year. Breeding in New Jersey started again in 1930 (9), and by 1955 this species had extended north to Island Beach.* Its breeding range has even expanded to include parts of coastal New England in recent years.

High counts include 400 at North Arlington (September 20, 1976), more than 1,000 at Stone Harbor (summer 1966), and an impressive total of 3,035 with a coastal New Jersey survey in 1979.

LITTLE BLUE HERON *Egretta caerulea* (Linnaeus)

The Little Blue Heron is a summer resident from mid-March through November, mostly along the coast. New Jersey was formerly the northern limit of its breeding range, but it has expanded to Long Island, New York, and beyond. Nesting occurs from May to June, and like many other species of herons and egrets it has postbreeding dispersal in August that distributes the bird widely throughout the state. (The majority of the wanderers are white-plumaged immatures which might be mistaken for egrets.) The nesting history in our state starts with old records from the 1780s, but like other herons and egrets these too declined so that none was breeding here by the turn of the century. In fact, modern breeding records did not occur until 1935, with 10 pair near Camden.

High counts include 240 mature birds at Stone Harbor (August 1966), and 353 in a coastal New Jersey survey in 1979.

TRICOLORED or LOUISIANA HERON *Egretta tricolor* (Müller)

The Tricolored Heron is an uncommon summer resident, mostly along the southern coast. Present from March to November, it nests in heron colonies on our coastal islands.

This southern species was first recorded in New Jersey at Cape May Point in 1920. More state records of individual birds came in 1926, 1929, 1930, 1932, and 1933. The first breeding record was in 1951 from the Stone Harbor heronry. By 1958 there were 210 nesting in three colonies (southern New Jersey). Like many other species in its family, this heron has continued its northward expansion of breeding to the New England coast. We find fluctuations in numbers: for example, a maximum of 397 was tallied at Stone Harbor in 1966, yet a 1979 survey of most of the New Jersey coast yielded only 193. In 1981 Stone Harbor tallied 1,000 for a new high count.

CATTLE EGRET *Bubulcus ibis* (Linnaeus)

Since the late 1960s this terrestrial egret has been a common summer resident (March through November) along our southern coast. It is irregularly reported at inland localities throughout the state, especially during spring migration (early April) and postbreeding dispersal (August). Very rarely, individuals remain into the winter months. The breeding season extends from April through July. It nests colonially with other species of herons and egrets in coastal rookeries, especially on undisturbed small islands with shrubbery and low trees.

This interesting addition to our state's avifauna arrived after a remarkable natural expansion of the species range from Africa to northern South America (about 1910) and then northward to the southeastern United States. It first appeared in Cape May on May 25, 1952. By 1958 it was being sighted regularly, and the first New Jersey nest was found on July 4, 1958, at Cape May. By 1960 there were 20 pair at Stone Harbor heronry, and the species was conspicuously spreading on New Jersey's southern coast. The northern part of the state had a major influx in 1962. A high count of breeding birds was tallied at Stone Harbor in August 1966. It is probable that the initial population explosion has now (1982) abated, although it still outnumbers all our native herons and egrets (Burger 1981).

GREEN-BACKED or GREEN HERON *Butorides striatus* (Linnaeus)

This little heron is widespread and common throughout most of the eastern United States. In New Jersey it is a summer resident from April through October at many ponds, swamps, and marshes.

Rarely, individuals remain into winter months at locations where open water remains. It nests throughout the state, from May to July. (It is usually more solitary in its nesting habits than other heron species.)

High counts include 110 (summer of 1974) and 150 (summer of 1973), both at the Great Swamp National Wildlife Refuge.*

BLACK-CROWNED NIGHT-HERON
Nycticorax nycticorax (Linnaeus)

This cosmopolitan species can be found in New Jersey through the year, but it is decidedly scarce in winter. Both it and the Yellow-crowned Night-Heron are most active just after sunset, but both do forage in the daytime as well. In our area it nests mostly on coastal salt-marsh islets, from May to July. (In other areas of its range it will nest in a wide variety of habitats, but always in colonies.)

High counts include 129 at Stone Harbor in August 1966, and 1,036 along the New Jersey coast in a 1979 survey.

YELLOW-CROWNED NIGHT-HERON
Nycticorax violaceus (Linnaeus)

This southern species is an uncommon summer resident along and near the coast, from April into October. There are various inland and northern New Jersey records (e.g. 1 at the Great Swamp* on May 10, 1980). It nests locally from May to June, in colonies with or near other herons.

This species was first recorded in the state at Woodbine in May 1892. Nesting was first reported from coastal Cape May County with 2 nests in the summer of 1927. High counts include 26 at Stone Harbor in 1966, and 155 in a coastal survey of New Jersey in 1979. It occurs in more than a dozen rookeries, but its future seems uncertain as more than half of the population is on barrier islands that are threatened by development. It is most likely to be seen away from nesting areas during the late summer, with postbreeding dispersal.

IBISES AND SPOONBILLS
Threskiornithidae

WHITE IBIS *Eudocimus albus* (Linnaeus)

Although the White Ibis breeds as near as North Carolina, it wanders to New Jersey only rarely as a summer visitor. Records from the mid-1800s were mentioned by Audubon and Turnbull. The first modern state record comes from Stone Harbor, where an immature bird appeared on August 17, 1951 (2 were here in 1958). More coastal sightings came in 1962, 1965, 1966, and 1968. An increasing number of records came in 1972, 1975 (a mild irruption), and 1977 with more than 30 in the state. (In recent years there have also been numerous records from neighboring states.) Rarely, individuals will remain in southern New Jersey into the fall, and one stayed into winter at Cape May (February 10, 1978). Although the majority of sightings come from the coast there are inland records (notably with the 1977 invasion) and observations along the Delaware River as far north as Scudder's Falls* (1980). Most reports involve immature birds, but there have been 3 adults over the years at Stone Harbor (most recently from April to October 1981).

GLOSSY IBIS *Plegadis falcinellus* (Linnaeus)

The Glossy Ibis is a common summer resident along our southern salt marshes. It nests from April to July in coastal colonies with herons and egrets. Its New Jersey status changed dramatically from very rare to common within the period 1955–1975. This increase was just part of a large regional expansion, and the species now breeds as far north as Maine, although reproductive success is often lower in northern areas.

The first state record came from Great Egg Harbor* on May 7, 1817 (the bird was collected). In 1932 there was a report from the Metedeconk River,* and 4 were sighted at Paulsboro in 1937. Stone Harbor had 3 appear in 1952 with possible nesting. The first definite nesting occurred here in 1955, and the next year Brigantine National Wildlife Refuge* had 4 pair. By 1958 Stone Harbor had 122 and was increasing rapidly with a count of 916 by 1966 (about 1,700 were tallied here in 1981). And there are lesser numbers in other heronries along the coast, but this species shows considerable variation in numbers from year to year. Perhaps we may soon attain stable populations below those of the peak years mentioned above.

WHITE-FACED IBIS *Plegadis chihi* (Vieillot)

This accidental visitor comes to us from western parts of the United States. New Jersey has a set of sightings that may only involve one bird: Brigantine National Wildlife Refuge* on April 17, 1977, and nearby Tuckerton on April 21 of the same year. There was also a report from Brigantine Refuge* in July–August 1983. (There are several old New York State records.)

STORKS / *Ciconiidae*

WOOD STORK *Mycteria americana* Linnaeus

In New Jersey the stork is but an accidental visitor from southern cypress swamps. Records include 1 at Cape May in August 1922, 4 there in July–August 1923, 1 at Stone Harbor on September 23, 1951, a Cape May Point bird on November 13, 1977, 1 at Leisure Town on March 8, 1978, an immature at Cape May Point on June 12, 1979, and 2 at Cape May Meadows* in September 1983. Curiously, both Pennsylvania and New York have had more records of this species, but it is still considered accidental in any northeastern state. (The Wood Stork is a permanent resident of Florida, but it has considerable postbreeding dispersal when it will soar northward for hundreds of miles.)

FLAMINGOS / *Phoenicopteridae*

GREATER FLAMINGO *Phoenicopterus ruber* Linnaeus

Although some records of this tropical species could be wandering wild birds from the Caribbean, it is most likely that New Jersey's sightings usually involve birds recently escaped from captivity, either locally or from Florida, with subsequent wandering.

Reports include 1 at Barnegat Light (October 17, 1964), 2 in flight near Atlantic City (October 26, 1976), and a 1979 bird at Sea Bright (the species of this flamingo was questioned). Most interesting, however, was a summering bird at Brigantine National Wildlife Refuge* in August 1964, a year when there were several sightings in northeastern states.

DUCKS, GEESE, AND SWANS *Anatidae*

FULVOUS WHISTLING-DUCK *Dendrocygna bicolor* (Vieillot)

In recent years this duck has been a rare visitor to New Jersey, from southern states. It nests in coastal marshes north to North Carolina and was first recorded in New Jersey in 1961. Today it is seen almost annually in small numbers. Sightings include a flock of 6 at Brigantine National Wildlife Refuge* (October 28–29, 1961, seen by hundreds of observers), birds at the Trenton Marshes* (April–May) and Cape May (May) in 1965, and birds at the Secaucus Marshes* (May) and Brigantine National Wildlife Refuge* (fall) in 1966. More recently there was a large flight of up to 40 individuals in the state in the spring of 1974, and a smaller "echo" flight in 1975.

TUNDRA or WHISTLING SWAN
Cygnus columbianus (Ord)

New Jersey lies just to the north of the main wintering area (Chesapeake Bay) of this arctic nesting species. In our state it is an uncommon or rare transient (March and November); it has increased both as a migrant and as a winter resident since 1960. Still, only small numbers are present during the winter (e.g. usually less than a total of 200 on all New Jersey Christmas counts).

High counts include 387 in the Pine Barrens* bogs (March 26, 1978), 500 at Whitesbog (March 4, 1980), and 1,000 at the Elmer Christmas count (January 3, 1981).

MUTE SWAN *Cygnus olor* (Gmelin)

This bird, largest of the world's waterfowl, is a permanent resident, primarily along the coast and on northern lakes. It nests from March through June, in loose colonies amid aquatic vegetation of lake borders. Many small populations throughout the state are semidomesticated (e.g. pairs are often introduced to park ponds).

This species was brought to North America from Europe; the first New Jersey record was at Elizabeth, on October 24, 1916. By 1920 the species had become established on coastal ponds from Asbury Park to Point Pleasant. High counts include 83 on a Christmas count in Sussex County (December 16, 1979), 364 at Lakehurst on a Christmas count (1978–1979), and about 350 on Barnegat Bay* during the winter of 1980–1981.

LESSER WHITE-FRONTED GOOSE
Anser erythropus (Linnaeus)

This European species can be listed only hypothetically for New Jersey. A recent record (Smithville Pond,* September 1971) may have involved an escaped bird, particularly as this species is frequently kept in captivity. Adjacent states (e.g. Pennsylvania) have had a few recent records that are considered to involve wild birds.

GREATER WHITE-FRONTED GOOSE

Anser albifrons (Scopoli)

This medium-sized goose is a very rare winter visitor from the West (there is a much smaller possibility of individuals coming to our region from nesting areas in Greenland). Most sightings come from the coast, in November.

Reports include 1 at Barnegat Bay* on November 28, 1926, 2 at the bay on November 10–11, 1928, and 4 there on November 10, 1935. A state high occurred with 5 at Beach Haven Inlet* on November 16, 1945. Recently birds were reported at eastern Monmouth County (September 30, 1953), Brigantine National Wildlife Refuge* (winter of 1957–1958, April 1961, winter of 1974–1975, and January and May 1967), Salem (November 11, 1958), Bridgeton (1966 Christmas count), and Plainsboro (April 16, 1972), with 2 at Moores Beach (May 23, 1981) and Lawrenceville (December 19, 1981).

SNOW GOOSE

Chen caerulescens (Linnaeus)

The Snow Goose is a common migrant and a winter resident in southern parts of the state. The species has extended its period of wintering in recent decades with some flocks remaining even into May. With prolonged periods of residence its grazing can seriously damage local vegetation (e.g. the salt marshes at Brigantine Refuge*). During migration it can be seen passing overhead anywhere in the state, but it is decidedly rare away from the coast.

High counts include 2,000 on the Raritan Bay,* December 13, 1948, and a remarkable 20,000 at Heislerville in late February 1976. A statewide survey reported over 70,000 in December 1980. It is an important species for our coastal hunters.

With current taxonomy this species includes a bluish gray phase that was formerly considered a distinct species, the "Blue Goose." In our area the Blue Goose is quite rare, but it has been increasing slightly in frequency in recent decades. A high count of this dark form was 12 at Heislerville on March 20, 1979 (with thousands of birds of the white form).

ROSS' GOOSE

Chen rossii (Cassin)

The Ross' Goose is an accidental visitor from the West; it is very rarely found along the coast of the Gulf and Atlantic states. There are only a few New Jersey records: Brigantine National Wildlife Refuge,* an adult from January to April 1972, an immature April 1–23, 1972, 1 November 16, 1976, and 1 late October–December 1982 (excellent photographs of this bird were obtained). This small goose often shows a strong association with Snow Geese, with which it, in fact, sometimes hybridizes.

BRANT *Branta bernicla* (Linnaeus)

The Brant is an abundant migrant and winter resident along the shore. Its migration peaks in March and November with flocks of hundreds passing our coast. Summering birds may sometimes total in the thousands (e.g. in 1953 and 1954), but there is no nesting south of arctic Canada.

High counts often come from the area of Atlantic City (e.g. 75,000 at Brigantine Refuge,* October 20–26, 1962). Inland reports are scarce except of migrants along our mountain ridges. The Winter Waterfowl Trend Surveys (U.S. Fish and Wildlife Service) show good numbers of Brant along the New Jersey coast in recent winters (e.g. 53,605 in 1980–1981, and 63,000 in 1981–1982), although they may still be below the high populations of the 1950s.

The former "Black Brant," *Branta nigricans,* is now included as a form of this species. This blacker race is western, but it has been recorded in New Jersey several times including a specimen from Egg Harbor (1846), Brigantine Island* (December 4–18, 1976, many observers), and Nummy Island* (January 19, 1979).

BARNACLE GOOSE *Branta leucopsis* (Bechstein)

This species is considered hypothetical for New Jersey; the only specimen record of this European goose, collected in Bergen County, March 21, 1926, is of a bird suspected to be an escape. More recent sightings are also believed to probably involve escapes of this species which is frequent in waterfowl collections. Furthermore, very few reports from either Pennsylvania (Wood 1979) or New York (Bull 1974) are considered to involve truly wild birds.

CANADA GOOSE *Branta canadensis* (Linnaeus)

This large goose is an abundant migrant and common winter resident. In recent decades the species has remained in small numbers through the summer, with some nesting. Winter populations have been high with recent mild years. Migration peaks from March to April and October to November, with flocks passing over the entire state.

It is difficult to determine when this species began nesting in New Jersey as a "wild bird"; breeding probably started here about 1944. Before then summering birds were park releases, feral birds, or of otherwise questionable status. Brigantine National Wildlife Refuge* had 5 pair in 1959, and today the refuge is one of our main breeding locales. Winter populations can be large enough (e.g. 12,318 at Princeton in December 1981) to produce local grazing damage to lawns. The Canada Goose and Snow Goose are both very important to New Jersey's goose hunters (a combined total of more than 21,000 was taken in 1978).

The high count appears to be an Island Beach* record of 50,000 passing by on October 20, 1967. The very small race, *B. c. hutchinsii,* has been noted rarely within coastal flocks of the larger Canada Geese (e.g. April 16–30, 1972, at Brigantine National Wildlife Refuge*).

WOOD DUCK *Aix sponsa* (Linnaeus)

This attractive waterfowl is a migrant and summer resident in many freshwater habitats of our state. Small numbers remain in winter, usually in our southern and coastal areas. Migration is not pronounced but peaks in March and September–October. Nesting continues in the period of mid-March through June, in wet wooded areas (tree cavities are the natural nest sites). Populations were precariously low at the turn of the century, reportedly from overhunting. The species has recently responded well to nest-box management.

The Great Swamp National Wildlife Refuge* has had spectacular success with Wood Ducks. In the summer of 1973 the refuge had 4,700, with 95 percent hatch at 400 nesting boxes. In 1977 there were 750 nest boxes used, producing 1,231 ducklings. And the summer of 1979 gave a spectacular high count of 17,000, including 2,000 young.

GREEN-WINGED TEAL *Anas crecca* Linnaeus

This small freshwater duck is a common migrant and winter resident. A few are present in summer, with limited nesting records.

Migration peaks March–April and October–November. (Fall movement starts as early as late August.)

The first state nesting report was in 1960 at Brigantine National Wildlife Refuge.* (Hackensack Meadows* had clutches that did not hatch in 1962.) An estimate of 280 at Great Swamp* in June 1974 strongly suggests breeding, and there is a recent nesting record from Whitesbog.

A high count of 5,000 was tallied at Brigantine National Wildlife Refuge* in late fall 1967. The Green-winged Teal is an important New Jersey game species, with more taken in the hunting season than any other duck except the Mallard and Black Duck.

The Eurasian race of this species, formerly called the "Common Teal," is rarely reported. It is usually found within flocks of our American Green-winged Teals during spring migration (April).

AMERICAN BLACK DUCK *Anas rubripes* Brewster

The Black Duck has been the most abundant of the eastern freshwater ducks for many decades. In New Jersey it is a permanent resident, but most numerous in winter and during migration. It nests in a wide variety of marshes and ponds from March into August. It is still the most common species in the fall harvest of New Jersey duck hunters (e.g. more than 37,000 bagged in 1978). There is concern, however, about genetic mixing by breeding with the Mallard, a species that is rapidly increasing in the East. In the future the Black Duck, a truly characteristic resident of marshes along our coast, may decline because of interaction with more Mallards. Such a decline would be felt by hunters and birders alike.

A winter survey of waterfowl along the coast yielded a maximum of 60,000 Black Ducks, although the number can vary considerably from year to year.

MALLARD *Anas platyrhynchos* (Linnaeus)

Common as a migrant and winter resident, the Mallard includes mixed wild and domestic stock in New Jersey. Certainly the smaller numbers in summer contain some nesting descendants of captive stock. In the wild the species is increasing throughout much of eastern North America, and our yearly counts continue to rise. It is primarily a freshwater duck, nesting from mid-March into July. The nests are often in vegetation at lake edges, but they can also be located in a wide variety of upland sites (e.g. agricultural fields).

High counts include 4,000 at the Great Swamp National Wildlife Refuge* on October 1, 1971, and 6,000 there on November 30, 1972.

NORTHERN PINTAIL *Anas acuta* Linnaeus

The Pintail is a common migrant and winter resident. Migration peaks on inland lakes and ponds during February and early March, and November. There are no New Jersey breeding records for this species, but it has nested in New York and Pennsylvania (as nearby as Tinicum Refuge in Philadelphia).

High counts include 1,400 at the Great Swamp National Wildlife Refuge* (March 12–18, 1971), 3,000 at Troy Meadows* (March 15, 1942), 4,000 at Mad Horse Creek Wildlife Management Area* (March 15, 1980), and 5,500 at Pedricktown Marshes* (March 14, 1977). Note how all of these counts centered on mid-March migration.

BLUE-WINGED TEAL *Anas discors* Linnaeus

This shy species is a very common migrant in early April to May and September to mid-October. It is rare in winter, with only a few Christmas count sightings each year. Although it is uncommon as a summer resident, it has become a rare local breeding species in a few areas. For example, Brigantine National Wildlife Refuge* had 2 nesting pair in 1957 and 1960; Hackensack Meadows* had 20 pair (5 broods) in the summer of 1962.

A high count of 550 was reported from North Arlington on August 25, 1977.

CINNAMON TEAL *Anas cyanoptera* Vieillot

This western puddle-duck is an accidental visitor to our coastal marshes. There is a fair possibility that some records reflect escapes from captivity, but it is also likely that some wild birds do stray eastward occasionally. (Since the 1950s there have been scattered records from all of our neighboring states.)

There was an unconfirmed report of 3 shot by hunters in the Hackensack Meadows* (1958), and a well-substantiated sighting at Brigantine National Wildlife Refuge* in September 1962. Other reports come from the Brigantine Refuge* in the fall of 1963 and the end of November 1974.

NORTHERN SHOVELER *Anas clypeata* Linnaeus

The Shoveler is a migrant and winter visitor, with migration peaks from March to April and mid-September to mid-November. Formerly it did not range as far east as today, and it was considered rare in New Jersey before the 1930s. With modest but continual increases it became regular by 1950, the year of our first state nesting. Pairs nested at Brigantine National Wildlife Refuge* in 1959 and 1960, and today it is an uncommon local breeder at scattered localities. As yet there appear to be no permanent nesting populations in the state, however.

A high count of 2,500 comes from the Brigantine Refuge* in the late fall of 1967.

GADWALL *Anas strepera* Linnaeus

This western duck was formerly a rare visitor to the northeastern states. In recent decades, however, the breeding range has expanded eastward, and now it is an uncommon permanent resident in New Jersey. It can even be common during the migration and winter months. (A winter waterfowl survey tallied 1,500 in the state during the winter of 1981–1982.) Since its first nesting near Fortescue (July 1949) it has become one of our regular breeding waterfowl in the state. (The Atlantic coast breeding population extends from North Carolina to Rhode Island.)

Early high counts came from Brigantine National Wildlife Refuge* (e.g. 500 in 1955). This refuge also had nesting birds in the early years (e.g. 4 pair in 1960).

EURASIAN WIGEON *Anas penelope* Linnaeus

A rare European visitor, this wigeon is usually found near the coast in winter. Reports are invariably of males, as females are very difficult to distinguish from females of the American Wigeon. The first New Jersey record of the Eurasian Wigeon came in 1880, and over the past century it has become clear that single birds often return to the same locale for many successive winters.

A remarkable high count of 9 occurred on March 19, 1949, at Bidwell's Ditch* in Cape May County. More recently 2 drakes wintered with American Wigeons at Shark River* during the winter of 1980–1981.

AMERICAN WIGEON *Anas americana* Gmelin

Our native wigeon is a common migrant and winter resident. Migration peaks March to April and October to December. With an eastward expansion of its range this species has greatly increased as a New Jersey migrant since the turn of the century. (There are now scattered breeding records in neighboring states, and it will probably nest within our state in the future.)

A high count of 7,000 was reported from Brigantine National Wildlife Refuge* in the late fall of 1967.

CANVASBACK *Aythya valisineria* (Wilson)

This coastal duck is a winter resident and migrant, with migration peaks in March and November–December. The state's winter population of this species was notably low in the first part of this century, but today it is often recorded in good numbers from shore ponds and bays. A winter waterfowl survey tallied an estimated 22,300 in New Jersey during January 1982.

Counts of 5,000 were made on the Raritan Bay* on December 3, 1939, and at Newark Bay* on December 17, 1978. Another recent high count came from Bayonne with 6,000 on December 16, 1979.

REDHEAD *Aythya americana* (Eyton)

The Redhead is a migrant and coastal winter resident; migration peaks from March to mid-April and November to early December. The species suffered a significant decline about 1930, and is still uncommon in many years. The Redhead has been introduced as a breeding bird in numerous New York refuges with some success (e.g. nearby Jamaica Bay Refuge on Long Island, New York, had 70 young hatch in 1965).

Recent high counts have come from Barnegat Light, with 200 on January 12, 1981, and 350 in January 1977.

RING-NECKED DUCK *Aythya collaris* (Donovan)

The Ring-necked Duck is a migrant and winter resident. It is primarily found on bodies of fresh water, especially during migration peaks in March and November. There is no nesting in our state, but it has expanded its breeding range in the Northeast and is, for example, a regular nester at lakes in New York's Adirondacks. In our region this formerly scarce duck has gradually increased over the past several decades; generally, however, it remains uncommon.

Modern high counts come from Swartswood Lake* (300 on April 2, 1958), Mannington Marsh* (200 on March 10, 1980), and Assunpink Wildlife Management Area* (260 in mid-March, 1976, and 450 on March 19, 1982).

TUFTED DUCK *Aythya fuligula* (Linnaeus)

This coastal duck is an accidental visitor from Europe, where it is an increasing species. Some or even most of our New Jersey reports could involve escaped birds as it is popularly kept in captivity (e.g. in outdoor waterfowl collections). There have been only a few New York and Pennsylvania reports, and these have also been considered of doubtful origin.

The prime location for this bird in New Jersey has been at Edgewater, on the Hudson River. Records here include February 24 to April 1, 1966 (first for the state), November to December 1966, and again in December 1967. A high of 4 was at this location in the spring of 1977. There is also a February 1971 record from Point Pleasant.

GREATER SCAUP *Aythya marila* (Linnaeus)

The Greater Scaup is primarily a saltwater species, but it is also seen regularly on inland waters during migration. It can be a locally abundant winter resident along our coastal bays (e.g. Raritan Bay*). Migration peaks in March and November. (It breeds in the arctic, and there are no nesting records south of Canada.)

Large counts are often made in the vicinity of Sandy Hook, with enormous rafts of up to 50,000 birds. In maximum years aerial censuses of coastal waterfowl can tally 200,000 scaup in the state (mostly Greater Scaup). Another high count came from Brigantine Island* with 25,000 on March 27, 1977.

LESSER SCAUP *Aythya affinis* (Eyton)

This scaup is primarily a freshwater species, as contrasted with the more coastal Greater Scaup. As such, the Lesser Scaup is an uncommon migrant and winter resident on reservoirs, lakes, and bays throughout most of New Jersey. It is most numerous during spring migration, from mid-March to mid-April, and it is generally more frequent in the south-central and eastern portions of the state. It is difficult to get high counts of this species as it is often mixed in with large flocks of the numerically dominant Greater Scaup, and individual duck identification between these two close species can be quite tedious.

COMMON EIDER *Somateria mollissima* (Linnaeus)

This arctic seaduck is a rare winter visitor on our ocean front, usually near rock jetties. Today this species is the more commonly reported eider; in the past the King Eider was relatively more common. In New Jersey most birds are females or immature males. Very rarely there are summer records such as a second-year male at Holgate on June 27, 1981. (Summer birds are regular at Cape Cod, Massachusetts.)

A high count of 100 was made off Island Beach* on December 15, 1979.

KING EIDER *Somateria spectabilis* (Linnaeus)

Another arctic species, the King Eider is a rare winter visitant along the coast, usually at rock jetties. Often it is in mixed flocks or rafts with the Common Eider. Adult males are particularly infrequent in our waters, and the identification of females (and some immature males) can be difficult because of similarity with the Common Eider.

High counts include 14 at Long Branch (January 1957), and a remarkable 41 at Longport (February 23, 1971).

LABRADOR DUCK *Camptorhynchus labradorius* (Gmelin)

The status of this extinct duck must be hypothetical; apparently there are no substantiated New Jersey records for this species of the North Atlantic coast. In winter it occurred rarely but regularly south to the south shore of Long Island, New York (last taken there in the fall of 1875). Since it was also reported as far south as Maryland by Audubon, it seems most probable that individuals occasionally wandered into New Jersey's coastal waters.

HARLEQUIN DUCK

Histrionicus histrionicus (Linnaeus)

This northern duck is a rare coastal visitor in winter, usually at rock jetties and often with eiders. It has shown a modest increase in our area since the 1950s, and it is now recorded annually, albeit in small numbers.

High counts include 9 (1 male) at Cape May from January to March 1981, 6 at Barnegat Light on February 17, 1961, and 6 (4 males) at this same location on February 19, 1956. During the winter of 1977 there were at least a dozen records from southern New Jersey. An unprecedented inland record came from Lake Parsippany* (an adult male) on November 12, 1958.

OLDSQUAW *Clangula hyemalis* (Linnaeus)

The Oldsquaw nests in arctic areas as far north as northern Greenland. It comes south to coastal New Jersey as a migrant and winter resident. Migration peaks in March and November, when it is sometimes common. It is very infrequent inland (e.g. at larger reservoirs), and rarely there are summer individuals along the coast (sick birds?).

High counts are often made on Christmas counts (e.g. 534 at the Raritan Estuary* on December 30, 1979, and a remarkable 8,937 at Barnegat on January 4, 1981).

BLACK or COMMON SCOTER

Melanitta nigra (Linnaeus)

Most individuals of this species breed along the western and northern coasts of Alaska. It comes to New Jersey as a migrant and winter resident along the coast. Migration peaks from March to April, and from October into November. Formerly the least numerous of the three scoters, this species is now almost as common as the numerous Surf Scoter. The Black Scoter is the most likely to be found on inland bodies of water.

In addition to the high scoter counts mentioned under the next species there is a tally of 5,000 Black Scoters off Absecon Inlet* on October 14, 1972.

SURF SCOTER *Melanitta perspicillata* (Linnaeus)

Like the White-winged Scoter, this species is a Canadian breeder. In our state the Surf Scoter is a common migrant and winter resident along the coast. Its maritime migration peaks from March to mid-April and from mid-October through November.

High counts of scoters dominated by this species include 29,500 at Avalon on October 17, 1979, and an estimated 250,000 in the ocean north of Cape May, January–February 1968. (Black Scoters were also abundant in both of these reports.)

WHITE-WINGED SCOTER *Melanitta fusca* (Linnaeus)

This scoter is also an arctic nester, and it migrates to our coast from northwestern Canada. Both as a migrant and a winter resident it is rather restricted to the ocean; only rarely is it noted on inland lakes. During the first half of this century tens of thousands occurred in offshore flocks during migration each year (April and November), but in more recent years this has become the least common of the three scoter species.

COMMON GOLDENEYE *Bucephala clangula* (Linnaeus)

The Common Goldeneye nests along lakes and rivers in northern forests, south to the Adirondacks of New York. In New Jersey it is a common migrant and winter resident along the coast. During migration it can also be found on inland lakes and reservoirs. The Common Goldeneye's migration peaks in March and November.

High counts come from the Delaware River with an estimated 200–300 near Dingman's Ferry, Pennsylvania (January–February 1977), and 200 near Manunkachunk, Warren County (January 1982).

BARROW'S GOLDENEYE *Bucephala islandica* (Gmelin)

In eastern North America the Barrow's Goldeneye has a very limited breeding range (primarily in northern Labrador). Predictably, therefore, it is a very rare winter visitant along our New Jersey coast, usually in the company of Common Goldeneyes. Most reports are of males, as the winter female is difficult to distinguish from the female of the Common Goldeneye.

Earlier records include 2 from the Raritan Bay* (January 5, 1936, and February 5, 1939). With remarkable fidelity, one bird returned for at least 13 consecutive winters at Shark River (1969–1982). Possibly our state high count is 2, which occurred on the Barnegat Christmas count, January 4, 1981.

BUFFLEHEAD *Bucephala albeola* (Linnaeus)

The Bufflehead nests on woodland lakes and ponds, east to Ontario and Quebec. In our area it is a common migrant and winter resident, especially along coastal bays. During migration it is regularly seen at inland bodies of water (e.g. large reservoirs and the upper Delaware River). It is considered more numerous today than earlier in the century. In recent years we have had very rare summer records of a few nonbreeding individuals.

HOODED MERGANSER *Lophodytes cucullatus* (Linnaeus)

This attractive species is a migrant and winter visitant, inland and especially along the coast. It is reported very rarely in summer, but there are at least seven breeding records in the state (e.g. Cape May County in 1949, Warren County in 1962, Essex County in 1980, and Sussex County in 1980 and 1981). This cavity-nesting duck breeds very locally throughout much of upstate New York.

High counts come from Brigantine National Wildlife Refuge* with 400 in late December 1974, and 500 in the winter of 1980–1981.

COMMON MERGANSER *Mergus merganser* Linnaeus

In eastern North America the Common Merganser breeds on boreal lakes across Canada and the northern states. In New Jersey it is a winter resident and migrant, primarily at fresh water. This species may nest rarely along the northern Delaware River (e.g. adults with small young on the Big Flatbrook* tributary in the summer of 1978). It has bred in nearby New York (e.g. Catskills) and Pennsylvania (e.g. Luzerne County).

High counts include 400 at Spruce Run Reservoir* (March 31, 1978, and February 21, 1981), 400 at Split Rock Reservoir* (November 30, 1980), 1,200 at Point View Reservoir* (March 20, 1981), and 1,300 at Lake Assunpink* on March 9, 1982.

RED-BREASTED MERGANSER *Mergus serrator* Linnaeus

As a coastal migrant and winter resident, the Red-breasted Merganser is the most common of our three merganser species. Our area is at the southern limit of its breeding range, and it is very rarely reported as a summer breeding bird from both coastal Long Island, New York (e.g. 1971) and New Jersey (e.g. Barnegat Bay* in 1937 and 1950).

RUDDY DUCK *Oxyura jamaicensis* (Gmelin)

The Ruddy Duck is a common migrant and winter resident, favoring freshwater ponds and lakes. This species has increased in New Jersey in recent decades, and although its breeding range is mostly west of the Mississippi there have been scattered nesting records for most northeastern states since 1950. In the 1930s the species was rare along the coast, but by the 1950s there were winter counts of 40,000 or more on the Delaware River near Floodgates* (Gloucester County). Unfortunately, there has been a considerable decline in the last decade (major oil spills in the mid-1970s *may* have caused some mortality of this species). A recent high count of 17,280 was made on the river, November 16, 1976. The first nesting in our state was at the Hackensack Meadows* in 1958; 10 pair nested at Brigantine National Wildlife Refuge* in 1960.

MASKED DUCK *Oxyura dominica* (Linnaeus)

This tropical species is only an accidental visitor to northeastern states. A female was reported at the Brigantine National Wildlife Refuge* in the 1950s (date?), and there was an unconfirmed report of a flock of 7 on the northern New Jersey coast after Hurricane Donna on September 12, 1960. The species has accidentally occurred in other East Coast states as well (e.g. Maryland and Massachusetts).

AMERICAN VULTURES / *Cathartidae*

BLACK VULTURE *Coragyps atratus* (Bechstein)

This tropical species was formerly a very rare spring visitor; now it is reported annually in small numbers in any month of the year. Older records come from the period of spring dispersal (March–May). Wintering records mostly come from the southern half of the state, especially Salem County (e.g. 13 on the 1981–1982 Christmas count). The Black Vulture breeds in southern Pennsylvania, and recently nesting occurred in western New Jersey.

New Jersey records include 1877, 1930, 1934 (2), 1939, 1943, 1953, 1954 (2), 1961, 1962, 1964 (2), 1966, 1968, 1969, 1970 (2), 1972, 1973, and then annually with multiple records. High counts include 7 at Princeton on February 3, 1980, and 17 at Titusville in early December 1980. The first documented breeding in New Jersey came from a cave in Hunterdon County (2 nestlings in May 1981). Future nesting in other counties seems likely, because the species continues to increase gradually in our region. Curiously, the Black Vulture is having a prolonged decline in much of its southeastern range in the United States.

TURKEY VULTURE *Cathartes aura* (Linnaeus)

The Turkey Vulture is a permanent resident, but it generally retreats from or becomes uncommon in northern parts of New Jersey in mid-winter. Although this species has probably nested throughout the state we have few records, as the nest sites tend to be well concealed and in remote locations (e.g. forested areas of Morris and Sussex counties). The species has increased in the northeastern states in recent decades, even expanding into New England. It seems probable that this carrion feeder has benefited from the high frequency of mammalian roadkills along modern highways.

High counts include a remarkable 559 on a Cape May Christmas count (1946), 959 at Cape May during the fall migration season (1976), and 200 at a winter roost near Dennisville (January 5, 1969).

OSPREY, KITES, HAWKS, AND EAGLES
Accipitridae

OSPREY — *Pandion haliaetus* (Linnaeus)

The Osprey is a migrant and local summer resident, nesting along the coast and Delaware Bay shore. Migration peaks in April and September. Rarely, individuals remain into winter (e.g. in 1974 a total of 3 was seen on New Jersey Christmas counts).

Prior to 1950 there were about 500 active nesting pair in the state. After pesticide-related reproductive declines we had a low point of 50 nests in 1974. With the state's efforts in an egg transfer program and subsequent natural recovery the total number of nests rose to 70 in 1977, and 86–87 in 1979–1981. In the summer of 1981 there were an estimated 174 nesting adults, 89 fledglings, and some nonnesting adults. Most nests were in Cape May County (44) and Ocean County (15). Active nests were also reported in Atlantic, Monmouth, Burlington, Salem, and Cumberland counties. The enormous stick nests atop a dead tree or telephone pole are used year after year.

Seasonal high counts include 574 for the fall of 1981 at Upper Montclair, and 2,206 for Cape May in 1982. Daily highs include 91 at Upper Montclair (September 21, 1976), 265 and 228 respectively at Cape May (October 8, 1979 and October 3, 1982).

AMERICAN SWALLOW-TAILED KITE — *Elanoides forficatus* (Linnaeus)

This kite is a very rare visitor from southern states (today the species breeds no further north than South Carolina). Sightings became more frequent in our area in the 1970s. Almost all reports come in the May–June or September migration periods.

State sightings include Bordentown (November 1883), Cumberland County (June 4, 1893), Morris County (September 18, 1887), Raritan River* (May 19, 1940), and Cape May (September 1, 1946). More recently reports have come from Cape May (May 1962, May

12, 1972, spring 1976, May 30, 1978, May 24 and June 12, 1979, and several birds April–May 1980 and June 18, 1982), as well as Avalon and Sandy Hook (spring 1976), Swainton (June 24, 1975), Paterson (September 20, 1970), West Orange (May 4, 1974), and Greenbrook Sanctuary* (May–June 1982).

MISSISSIPPI KITE *Ictinia mississippiensis* (Wilson)

The Mississippi Kite has been a very rare visitor from southern states, but recently it has had a rapid range expansion toward the north. There is at least one report from the past century, but almost all records are fairly recent.

Cape May has been the principal site for sightings: May 30, 1924, June 3 and September 29, 1976, June 3, 1977, early June 1978 (possibly 2 birds), May 30, 1979 (2 subadults), May–June 1980 (at least 4 birds), August 1, 1980, and September 2, 1980 (2 birds). In 1981, between April 9 and June 15, probably 9 individuals were seen around Cape May with a maximum of 4 on June 7, 1981. This number in the breeding season suggests that we might have New Jersey nesting in the near future.

Other locations of Mississippi Kite reports are Sergeantsville (May 8, 1961), Port Norris (March 28, 1965), Canton (June 24, 1979), West Orange and Assunpink* (both in May 1980), and Sunrise Mountain,* Sussex County (October 29, 1980).

BALD EAGLE *Haliaeetus leucocephalus* (Linnaeus)

Our national bird is a rare New Jersey migrant and very rare winter resident, mostly near the Delaware River or the coast. Except for a single pair it has been extirpated as a breeding bird in our state. It is a notably early migrant in the fall (starting in late August). Statewide surveys in mid-winter usually find 6–8 birds (e.g. at Dingman's Ferry, Pennsylvania, on the Delaware River, and Dividing Creek in Cumberland County). At the turn of the century there were as many as 30 nesting pair, primarily in Cumberland and Salem counties. Prior to the 1960s the state had about 10 nesting pair, but by 1961 7 active nests contained only 1 young (probably because of pesticide interference in breeding). The first successful nesting since 1967 occurred in 1974 with 2 young raised in three southern New Jersey nesting attempts. The final nesting pair has not fledged any young of its own since 1976 (this pair continues to use a tall pond pine in Cumberland County). The state's Endangered Species Program has transferred young eaglets to be adopted at this nest (e.g. a pair in 1983). Also, other young eagles were released in 1983.

High counts include 17 at Raccoon Ridge* on September 11, 1939, and 10 (only 1 adult) at Brigantine National Wildlife Refuge* on December 31, 1960. More recently 8 were tallied in a survey of Cumberland County (December 29, 1979), and 8 (2 adults) were counted along the upper Delaware River (January 11, 1981). Thirty wintered in New Jersey in 1979–1980, and 22 (8 adults) in 1980–1981. The Cape May hawk lookout tallied 25 during the fall migration of 1981.

NORTHERN HARRIER *Circus cyaneus* (Linnaeus)

The Harrier is a migrant and winter resident. It is quite scarce as a breeding bird, primarily with ground nests in salt marshes along the Delaware Bay shore. (Formerly it was a more common nesting species.) Migration peaks in April and mid-October into November.

Various kinds of high counts include 71 banded in the fall of 1977 at Cape May, 110 on a Cape May Christmas count (1947), and 3,118 during the fall migration at Cape May. High seasonal counts at other hawk watches include 317 at Skyline Manor* (1979), 182 at Raccoon Ridge* (1978), and 214 at Upper Montclair (1981). A nesting area in the Delaware Bay shore had about 20 pair in 1978 and 34 pair in 1983. Winter roosts of this species are now rare in New Jersey; the largest recent one contained 14 birds at Alpha, Warren County in the winter of 1982. (In the 1950s roosts in Hunterdon County would hold more than 100.)

EUROPEAN SPARROWHAWK *Accipiter nisus* (Linnaeus)

This European raptor is strictly an accidental visitor. It was seen once at Cape May Point on October 24, 1978. If this detailed sighting is accepted it would be the first North American record. As its appearance is quite similar to that of our small *Accipiter* hawks some photographic confirmation would have been significant.

SHARP-SHINNED HAWK *Accipiter striatus* Vieillot

This small hawk is a common migrant, uncommon in winter, and scarce as a breeder in summer. (Apparently it was generally more numerous in the earlier decades of this century.) Nesting is rather restricted to coniferous forests in the northern part of our state, although there were a few local nestings in southern pine-oak areas until 1954. Summer birds were also seen in the cedar swamps of the Pine Barrens* in 1980.

The Sharp-shinned is a conspicuous fall migrant from mid-September into the last half of October. Cape May has truly spectacular flights of this species with seasonal counts of 32,610 (fall 1978), 48,447 (1979), 52,273 (1980), and 43,061 (1981). Daily high counts at Cape May include 4,608 on October 8, 1979, and 4,364 on September 25, 1978. During the fall season of 1979 a remarkable 3,539 were banded at Cape May.

COOPER'S HAWK *Accipiter cooperii* (Bonaparte)

The Cooper's Hawk is a migrant, uncommon or rare as a winter resident, and scarce as a breeder. This species declined in numbers over a period of many years but may now be increasing. Migration peaks in April and mid-September into November. There are some New Jersey nesting sites, mostly in the counties of Sussex, Warren, and Hunterdon. Its stick nest is usually high in a forest tree (deciduous or coniferous). The ratio of the very similar Sharp-shinned Hawks to Cooper's Hawks seen at the Raccoon Ridge* in fall migration can be as high as 40:1, so identifications should be cautious.

High counts again come from Cape May with 1,875 during the fall of 1979, and 2,086 in the fall migration of 1981. A daily high of 347 was made on October 18, 1979, at the Cape May hawk watch. A record number was also banded here, 189 in the fall of 1979.

NORTHERN GOSHAWK *Accipiter gentilis* (Linnaeus)

This powerful hawk of northern forests is a rare breeder, migrant, and winter visitant. In recent years there have been more breeding records (e.g. throughout undisturbed areas of north New Jersey), and 1982 saw 18 nesting in the state. Nests are placed quite high in tall trees, usually in heavily forested areas. (They show some propensity to nest near forest trails.)

A big flight of this species occurred in the fall of 1972 (Raccoon Ridge* tallied 202). Daily high counts of 13 were recorded at both Cape May (November 9, 1973) and Raccoon Ridge* (October 20, 1977, and November 16, 1969). Reports as far south as Cape May have been increasing in recent years: the seasonal total for the fall of 1981 there was 37, and the fall of 1982 yielded 41.

RED-SHOULDERED HAWK *Buteo lineatus* (Gmelin)

The Red-shouldered Hawk is an uncommon resident and a common migrant. Migration peaks in March and late October into early November. Formerly it was more widespread in its nesting, but today it is primarily limited to moist woodlands in northern New Jersey (e.g. Pequannock area). (Birds return to the same nest each year.)

A seasonal high count of 700 was made at Cape May during the fall of 1976. A daily high of 290 was made at Rutherford on March 18, 1944.

BROAD-WINGED HAWK *Buteo platypterus* (Vieillot)

For short periods the Broad-winged Hawk is an abundant or common migrant, particularly in the fall (with a pronounced peak in mid-September). As a summer resident the bird breeds locally throughout the state, and most commonly in the hillside forests of the northern counties.

Seasonal high counts include 20,468 at Skyline Manor* (fall of 1978), 25,530 at Raccoon Ridge* (fall 1975), and 22,342 at Upper Montclair (fall 1981). A daily high of 9,414 was tallied at Skyline on September 13, 1978, and an impressive 10,827 at Montclair on September 21, 1981. An unprecedented count of more than 15,000 was made near Blairstown on September 14, 1983.

SWAINSON'S HAWK *Buteo swainsoni* Bonaparte

The Swainson's Hawk breeds in western North America, and has a long migration south into South America for the winter. On this migration a few strays are very rare fall visitors to our state. Because of the intensive hawk-watching efforts it is now recorded almost annually in New Jersey.

A supposed specimen from New Jersey (1915) has been questioned as to the geographic information on the label (*Cassinia,* 1973, 54:31). The first modern record then comes from Oradell, October 22, 1947. Subsequent sightings include New Vernon (October 23, 1971), Sussex County (November 1, 1975), Livingston (September

12, 1978), and the Turtle Back Zoo* (an adult on April 23, 1981). The last is our only spring record. Cape May, however, has had many sightings including 1 banded on September 16, 1973, an immature banded on November 17, 1975, 2 seen in the fall of 1976, and 1 on November 19, 1978, September 23, 1980, and September 29, 1980. In the fall of 1981 7 were tallied on the Cape May hawk count, between September 11 and November 9. Both the light- and the (rare) dark-phased birds have been seen in New Jersey.

RED-TAILED HAWK *Buteo jamaicensis* (Gmelin)

Although the Red-tailed Hawk is common as a migrant and winter resident, it is scarce as a breeding bird. At various times, however, it has nested throughout the state. Nesting occurs in forests, with a large stick nest used year after year. It is a dominant species in the late fall migration of raptors, with its peak between the last half of October and early November.

Seasonal high counts include 2,761 at Raccoon Ridge* (fall 1975), 1,601 at Montclair (fall 1981), and 2,777 at Cape May (fall 1973). Daily highs include 400 at Cape May (November 8, 1972), 600 at Raccoon Ridge* (November 1, 1980), and a spectacular spring migration count of 1,000 at Yards Creek* (March 2, 1972).

FERRUGINOUS HAWK *Buteo regalis* (Gray)

As an accidental visitor from the western United States, this species must be carefully identified by observers well familiar with the variability of plumage in other *Buteo* species. The Ferruginous Hawk is not particularly common within its normal range, and state bird books give no records for either New York (Bull 1974) or Pennsylvania (Wood 1979).

There are several records for the state, but good confirmations are usually lacking. Three sightings come from Cape May (November 26, 1978, October 27, 1979, and November 26, 1981). There is also a Dennisville report from November 27, 1979.

ROUGH-LEGGED HAWK *Buteo lagopus* (Pontoppidan)

This arctic nesting raptor is a winter resident throughout New Jersey, but it is generally uncommon and local. It is most frequent along the coast, at salt marshes or at rodent-infested landfills. Both light- and dark-phased individuals occur in the state, but the latter is decidedly scarce.

High counts include 15 in the Sayreville area on January 31, 1981, and 30 at Manahawkin on December 12, 1980. A seasonal total of 20 for the fall migration at Raccoon Ridge* in 1974, and a regional invasion of 27 (13 of dark phase) in northwestern New Jersey during February 1978 are also noteworthy.

GOLDEN EAGLE *Aquila chrysaetos* (Linnaeus)

The Golden Eagle occurs throughout most of North America, and it breeds as close as New York's Adirondacks (rarely). In New Jersey, however, this magnificent bird of prey is chiefly a rare migrant. Most observations are along the Kittatinny Mountains* during the fall (October–November), but records come from any month between September and April. Occasionally there are resident wintering birds (e.g. along the Delaware, or the southern coast).

High counts include 19 for the fall of 1979 at Cape May, and 27 for the fall of 1973 at Raccoon Ridge.*

FALCONS / *Falconidae*

EURASIAN KESTREL *Falco tinnunculus* Linnaeus

This European species is only an accidental visitor to the United States. An immature male banded at Cape May on September 23, 1972, was only the second North American record (the first was in Massachusetts in 1887). Subsequently, there was a sight record of 1 or 2 females at Cape May on October 9 and 14, 1979 (details in *Cassinia,* 1979, 58:19).

AMERICAN KESTREL *Falco sparverius* Linnaeus

This little falcon is common as a permanent resident throughout the state. It nests in tree cavities (or nest boxes) near fields, and it is quite tolerant of human activity near either hunting or nesting areas.

High counts include 1,937 banded at Cape May in the fall of 1976, 5,000 at Cape May Point State Park on October 15, 1961, 19,815 for the 1979 fall Cape May migration, and 21,828 there during the fall of 1981. Also from Cape May comes a remarkable estimate of 24,875 in one day (October 15, 1970)!

MERLIN *Falco columbarius* Linnaeus

This falcon nests in northern coniferous forests, and in our area it is a migrant and very rare winter visitant (along the coast). It is most numerous in fall migration (mid-September to mid-October), but it is also seen with some frequency in the spring migration (April). Off-season reports should be carefully considered for possible misidentification with Kestrels.

High counts include 225 banded in the fall of 1976, and 1,180 for the fall of 1979 (maximum of 129 in one day) at Cape May. A recent new daily high of 206 was made at Cape May on October 1, 1982, and a new seasonal high of 1,675 was made there in the fall of 1982.

PEREGRINE FALCON *Falco peregrinus* Tunstall

This spectacular falcon is a New Jersey migrant, particularly along the coast in fall (peak flights in October). Formerly it was a rare nesting species on cliffs along the Hudson River, but then it was lost as a breeding bird in the first half of this century. Now state reintroduction programs (in cooperation with the Cornell Laboratory of Ornithology) have a few birds nesting on our coast (e.g. Brigantine Refuge,* Tuckerton). The reintroductions have involved parental birds raised at Cornell University, and in 1980 New Jersey had the first eggs hatching in the wild east of the Mississippi since 1950! That year there were 4 young in two Ocean County nests. By 1981 there were 4 pair of introduced Peregrine Falcons, and in 1983 4 pair raised 14 young along the New Jersey Coast.

Migration counts have been increasing in recent years. For example, the Montclair count had 4 in the fall of 1977 (the first records there since 1974). At Cape May 21 were banded in the fall of 1976, and 230 were tallied in the fall migration of 1979 (maximum counts of 21, on October 2 and 6). Cape May also had a remarkable fall in 1982 with 363 Peregrines (a peak of 51 on October 1).

GYRFALCON *Falco rusticolus* Linnaeus

The Gyrfalcon breeds in the high arctic, and even in winter it remains almost entirely within Canadian latitudes. It comes to New Jersey only as a very rare winter visitor, and for many years we have no reports. Records include the following birds, with phase color indicated in parentheses when known: Brigantine (black) on February 9, 1940, Walnut Valley on October 21, 1945, Branchville (black) on March 12, 1950, East Rutherford (black) on January 4, 1953, Long Branch (white) on January 29, 1958, Bearfort Tower* (white) on October 27, 1969, Brigantine Refuge* between December 1971 and April 18, 1972, with 2 birds on some dates, and an immature (gray) at the refuge on March 15, 1975. It is rarely reported along the Kittatinny Ridge* (e.g. dark-phased bird at Raccoon Ridge* on November 29, 1975). Most recently it was at Stone Harbor (January 28, 1980) and at Rifle Camp Park* on a very early September 29, 1980. Between 1940 and 1983 there have been 18 Gyrfalcons recorded in our state (*New Jersey Audubon,* 1983, 9(1):2).

PHEASANTS, GROUSE, TURKEY, AND QUAIL
Phasianidae

GRAY PARTRIDGE *Perdix perdix* (Linnaeus)

This Old World game bird was introduced into northern New Jersey around the turn of the century, but it did not survive long. (The species has been a successful introduction in many northern areas, including New York State and parts of Canada.)

At least some reproduction took place in New Jersey. For example, a covey was reported on a farm near Elizabeth in 1912. Our state, however, lacks the cold climate and large areas of grassland and grain crops necessary for the proliferation of the Gray Partridge.

RING-NECKED PHEASANT

Phasianus colchicus Linnaeus

Another game bird of the Old World, the Pheasant was introduced and widely distributed in New Jersey by the early part of this century. The species is adaptable to a variety of open habitats, but it favors old fields with cover. Males are polygamous; females tend clutches in ground nests of weed stalks and grasses. It is still heavily stocked to meet hunting demands. Recent introductions have also included the dark green race, known as the "Japanese Pheasant."

A winter roost near Plainsboro had a high count of 27 in one group on a 1959 Christmas count.

RUFFED GROUSE

Bonasa umbellus (Linnaeus)

A permanent resident of forested areas, the Ruffed Grouse has a patchy distribution in New Jersey. In the Northwest it occurs along brushy mountains, and in southern New Jersey it is found less commonly, as a resident in the Pine Barrens.* More rarely the species is also a resident in central New Jersey (e.g. nesting in several areas of Mercer County). The beautiful ground nest is well lined with leaves and grouse feathers. Like many grouse species throughout the world, this bird can exhibit marked population fluctuations from year to year.

High counts come from the Great Swamp* with an estimated 100 in the summer of 1972, and 90 in the summer of 1973.

GREATER PRAIRIE-CHICKEN

Tympanuchus cupido (Linnaeus)

The "Heath Hen," the eastern subspecies of the Greater Prairie-Chicken, is now extinct. It formerly extended from Massachusetts to Virginia. In New Jersey it was most common in the short pine and oak "plains areas" of the Pine Barrens.* It was extirpated from our state by 1870. (The last records of the species are from Martha's Vineyard in 1932.)

It has been suggested that the Greater Prairie-Chicken of the midwestern race might be introduced in New Jersey, but the species is declining in most areas and it would be difficult to obtain surplus individuals for an introduction effort.

WILD TURKEY *Meleagris gallopavo* Linnaeus

The Wild Turkey was a former game bird of the New Jersey woodlands, extirpated in the early decades of the 1800s. Since 1954 some reports have come of wild birds crossing into northwestern New Jersey from Pennsylvania and New York. The state also began restocking in 1977 with 22 birds. By 1980 the population was estimated at 300–500 birds (!), and the next year the total was about 1,000 in a 250-square-mile area. Thus 1981 saw a hunting season for Turkeys in New Jersey—the first in more than 80 years. The spring season allowed hunting permits (900) for gobblers only, and 71 were reported shot.

In 1980 there were also transplants of birds to southern New Jersey, with 17 released in Ocean County and 30 in Belleplain State Forest.* (These involved birds transported from the northwest corner of the state.) The natural population of Wild Turkey in the Pine Barrens* had been lost more than a century ago.

A high count was obtained on the Walnut Valley Christmas count, with 44 birds on December 20, 1980.

NORTHERN BOBWHITE or BOBWHITE QUAIL *Colinus virginianus* (Linnaeus)

This species of Quail is primarily found from the central and southern plains states to the Atlantic coast. In New Jersey it is much more common in the southern counties, on the sandy soils of the coastal plain. Because it is a game species restocking is important in many areas, especially near the northern limits of its range.

The Bobwhite's favored habitat includes brushy fields and edges of pine forest. Nesting occurs from May to September, with two broods possible. The nest is well concealed, usually in grassy tussocks of abandoned fields. Populations throughout the state fluctuate markedly, particularly with high mortality in severe winters. In general the species was formerly more common, and many factors, including loss of agricultural land and excessive hunting, were probably involved in its decline. Unfortunately, released birds long ago were sometimes of other subspecies (e.g. from Florida), and such stock could easily have yielded harmful genetic mixing.

Stone's *The birds of New Jersey* (1908, p. 149) included the interesting note that our Quail are somewhat migratory, and that "flights are noticed at Cape May toward the end of October." Nevertheless, where present, the Bobwhite is predominately a permanent resident, seen or heard the year round.

RAILS, GALLINULES, AND COOTS
Rallidae

YELLOW RAIL *Coturnicops noveboracensis* (Gmelin)

This migrant of freshwater marshes is quite secretive and apparently very rare. It is mostly encountered in late summer and fall, although there have been a few winter and spring records.

High counts have been made at the Great Swamp National Wildlife Refuge,* where meadow mowing flushed as many as 6 in midsummer (June 19, 1974). It is most unlikely that any breeding was involved in these summer records, in spite of published suggestions otherwise.

BLACK RAIL *Laterallus jamaicensis* (Gmelin)

The Black Rail is a very rare migrant and local breeder. It nests in some of our southeastern salt marshes, and formerly it bred at wet grassy areas on barrier beaches (e.g. near Holgate). It was apparently more numerous in the 1920s and 1930s, when many nesting records were obtained in Cape May County. A few inland records have come during migration (e.g. Troy Meadows,* and Assunpink Wildlife Area*).

In recent years high counts have been made of birds calling at night (e.g. 10 in mid-June near Marmora, and other June counts at Manahawkin and the bay shore of Cumberland County).

CORN CRAKE *Crex crex* (Linnaeus)

This European species has been an accidental visitor to North America, but it is decreasing in its native land and we may expect few if any in the future. There are two old New Jersey records

(Salem, 1854, and Dennisville, 1905), and one recent single-observer sighting from Brigantine National Wildlife Refuge* (August 24, 1965). Similarly, 4 out of 5 Corn Crake specimens from Long Island, New York, were collected before this century (Bull 1974).

CLAPPER RAIL *Rallus longirostris* Boddaert

The Clapper Rail is a common summer resident of saltwater marshes, from mid-April to mid-October. It nests amid the marsh grasses, and eggs are easily lost if there are spring flood tides. Small numbers remain along our southern coast in winter.

The "mud hen" of hunters was taken in remarkably large numbers in the past (e.g. 10,000 killed in September 1896). In addition, eggs were collected as a food source, which was important in decreasing its numbers. In recent decades the species decline is probably related to widespread salt-marsh drainage.

KING RAIL *Rallus elegans* Audubon

New Jersey is at the northern limit of the King Rail's breeding and wintering range. For this reason it is generally a rare migrant and very local breeder in our freshwater marshes. Irregularly, there are coastal records, and a few are recorded during winter months in marshes with open water (especially at springs).

A high count of 50 was estimated at the Great Swamp National Wildlife Refuge* in the summer of 1973.

VIRGINIA RAIL *Rallus limicola* Vieillot

The Virginia Rail is an uncommon migrant and local breeder, mostly in freshwater marshes. Infrequently it will use brackish coastal marshes for nesting. At breeding areas birds are usually heard rather than seen, and this rail is easily attracted to tape recordings of its call. A few remain at open water in winter (e.g. at the Charles Rogers Refuge* in Princeton).

High counts include estimates of 250 at the Great Swamp National Wildlife Refuge* in the summer of 1973, and a staggering 500 there on October 8, 1972.

SORA *Porzana carolina* (Linnaeus)

This little rail is a migrant and summer resident in freshwater marshes. It will often build its nest *over* water, suspended in a cattail basket. Most of its breeding marshes are located in the northern and western portions of the state. Migration peaks from mid-April to mid-May and in late September. It is very rarely reported in the winter.

A high count estimate comes from the Great Swamp National Wildlife Refuge* with 200 in the summer of 1974.

PURPLE GALLINULE *Porphyrula martinica* (Linnaeus)

This attractive marsh bird is a very rare visitant from southern states. It is usually seen in spring or summer, with most records coming from southern New Jersey. Its breeding range is primarily within the Gulf states, but it has nested as nearby as Maryland and Delaware (e.g. in 1975). Summering birds were at Lily Lake,* Cape May in 1979 and 1981, and New Jersey breeding is a future, if unlikely, possibility. (During the summer of 1964 there were multiple sightings from May to July at Brigantine National Wildlife Refuge,* but again with no suggestion of nesting.)

There are more than a dozen records before 1950 including May 1892, May 1898, May 1902, May 1907, May 1928, May 1932, and August 1935. More current are one at Avon in June 1972 and one at Eatontown in April 1973. In 1974 there were four coastal records between April and June.

COMMON MOORHEN or COMMON GALLINULE *Gallinula chloropus* (Linnaeus)

The Common Gallinule is a migrant and local summer resident throughout most of the eastern United States, including New Jersey. It favors cattail marshes and has scattered nesting records throughout the state. It becomes decidedly uncommon in winter, although a few marshes provide Christmas count records (e.g. Trenton Marshes*).

High counts include 90 at the Mannington Marsh* on September 11, 1979, an estimated 150 at the Great Swamp* during the summer of 1973, and about 200 pair at the Hackensack Meadows* in the summer of 1962.

AMERICAN COOT *Fulica americana* Gmelin

Another pond and marsh species, the Coot is a common migrant and winter visitant, especially along the coast. New Jersey has some of the few nesting populations of this species near the Atlantic coast. It was formerly a rare nesting species, and then it bred regularly at larger freshwater marshes. Most recently, however, there have been substantial declines in most areas.

High counts include 300 pair at Hackensack Meadows* in the summer of 1962, and about 800 pair in Kearny Marsh* in the summer of 1980.

CRANES / *Gruidae*

SANDHILL CRANE *Grus canadensis* (Linnaeus)

This large bird comes to New Jersey as a vagrant from western North America, and it is very rare at any time in our region. Recently there have been a few state records each year, often of birds seen overhead (on migration). Sometimes individuals remain at a locality for months, foraging in meadows or marshes (e.g. a winter bird in Cumberland County in recent winters).

Records include 3 at Beesley Point, Cape May County, in 1857 (first state record), 2 at Cape May in October 1958, and many records of resident birds over a period of months at Stillwater, Yellow Frame, and Troy Meadows* from 1969 to 1972. Locations of more recent reports include Great Swamp* (May 1976), Raccoon Ridge* (March 27, 1976), Newton (March 29, 1977), Brigantine Refuge* (April 5, 1970, October 13, 1979, and May 10, 1980), and Cape May Point (December 5, 1980, and October 15, 1982).

WHOOPING CRANE *Grus americana* (Linnaeus)

Hypothetical. At one time this species was much more widely distributed, and it probably occurred along the Atlantic coast, including New Jersey, during migration. (It had been rarely recorded in

Pennsylvania prior to 1862.) In modern times it is one of North America's rarest birds and restricted to areas well west of the Mississippi.

PLOVERS / *Charadriidae*

BLACK-BELLIED PLOVER

Pluvialis squatarola (Linnaeus)

This species nests on the high arctic tundra. In our area it is a common coastal migrant with peaks of abundance in May and from mid-August through September. It is uncommon in winter, or in summer (nonbreeding individuals). On our coast it prefers beaches and salt-marsh flats, in all seasons.

High counts include 450 from a beach at Stone Harbor in mid-February 1971, and 1,500 at Brigantine Refuge* on August 19, 1981.

GREATER GOLDEN-PLOVER

Pluvialis apricaria (Linnaeus)

This European species breeds as near as Iceland and has been recorded on a few occasions in eastern North America. A possible record of this accidental visitor came from several observers at Cape May Meadows* on July 10, 1982 (*American Birds,* 1982, 37:164).

LESSER GOLDEN-PLOVER *Pluvialis dominica* (Müller)

This tundra nester is a fall migrant at short-grass habitats, such as sod farms in central New Jersey (the same lawns used by migrant Upland Sandpipers and Buff-breasted Sandpipers). It can also be found less frequently at salt-marsh flats, in association with other plovers. It is quite rare as a spring migrant as its main migration route northward goes to our west, along the Mississippi River states.

High counts include 100 at Cranbury on September 3, 1968, and 200 at North Arlington September 27–29, 1973.

WILSON'S PLOVER *Charadrius wilsonia* Ord

This plover of southern beaches regularly breeds as far north as Delaware, and formerly it was a regular summer resident along the coast of southern New Jersey. (It was most frequent at Cape May County, in the early nineteenth century.) Nesting became irregular in New Jersey after 1940 (last areas at Beach Haven Inlet* and Brigantine Beach*). By the mid-1960s it was extirpated as a breeder in the state, and today it is only a very rare visitant, most likely to be encountered during migration (e.g. 3 in April–May 1983).

Recent records include one at Brigantine Refuge* on May 26, 1976, and another at Cape May Point on September 28, 1979.

SEMIPALMATED PLOVER *Charadrius semipalmatus* Bonaparte

The Semipalmated Plover breeds in northern Canada and Alaska. In our area it is a common migrant along the coast with peaks in May and from August into early September. A few nonbreeding birds remain in summer, and it is generally rare in winter. Inland it is an uncommon migrant at lakeshore and reservoir edges.

A spectacular count of 5,000 was reported from North Arlington on July 26, 1973.

PIPING PLOVER *Charadrius melodus* Ord

The Piping Plover is a migrant and local summer resident, breeding on the outer coast. Migration peaks in March and August. Nesting colonies often have precariously low populations, and it is only tenuously established at many of its breeding areas, both in New Jersey and elsewhere.

It was reportedly common in Alexander Wilson's time (1812), and bred commonly in the state until 1890. The first nesting in this century occurred at Barnegat Light and Brigantine Beach* in 1921. The next year there were 5 pair nesting in the state, and by 1923 there were 10 pair. Before the end of the 1920s breeding was reported from at least seven other shore localities. Numbers of nesting pairs gradually increased through the 1970s, with a maximum of about 80 pair, most of which were in Cape May County (on both the Atlantic coast and the Delaware Bay shore). In 1982 the largest colony had 40 birds at Holgate. The flat beach areas used by nesting Piping Plovers are often shared with the Least Tern, another endangered breeding species.

KILLDEER *Charadrius vociferus* Linnaeus

This noisy plover is a common migrant and summer resident throughout. Unlike most other shorebirds, the Killdeer is content to nest far from the coast in a wide variety of upland habitats. Its four eggs are laid on a ground scrape in open habitats (e.g. lawn, field, or even pebble flat roof). Some winter over, particularly in southern New Jersey along the coast. Usually several hundred are tallied in the state on Christmas counts.

A high count of 1,000 was recorded at Mount Holly in October 1979.

OYSTERCATCHERS *Haematopodidae*

AMERICAN OYSTERCATCHER *Haematopus palliatus* Temminck

This is a coastal resident from March to November, nesting on barrier beach islands where there is little human disturbance. It is most common from May to August, but there are even some winter records.

At the time of Alexander Wilson (ca. 1812) there was a small breeding population within the state. It declined during the last part of the nineteenth century, and in the 1930s there were no New Jersey nestings. In the 1940s, however, small groups appeared near Tuckerton, with nesting in 1947. (In 1954 11 nests were located along the coast.) Since then it seems to have increased modestly, and as a species it is continuing to expand its range slowly northward past Long Island, New York, to the New England coast.

High counts include a remarkable 300 on Brigantine Island* on August 27, 1977, and 262 along the coast in a 1979 survey.

AMERICAN BLACK OYSTERCATCHER

Haematopus bachmani Audubon

This species is uncommon even in its native range on the Pacific coast. For New Jersey it is an accidental, with a single record. One was seen at Nummy Island,* near Stone Harbor, on August 29, 1951, by three observers. There are apparently no other East Coast sightings of this distinctive species.

AVOCETS AND STILTS
Recurvirostridae

BLACK-NECKED STILT

Himantopus mexicanus (Müller)

The Black-necked Stilt is essentially a tropical species that has only scattered nesting colonies north of Florida. It was formerly a very rare breeder in our southern counties as a coastal summer resident, through the early part of the nineteenth century. Now it is a very rare visitant from the South, usually in spring migration (April to May). Since the Stilt breeds as close as Little Creek, Delaware, it may return to New Jersey as a nesting species in the future. Reintroduction of nesting birds has also been attempted with egg transplants (put into Willet nests). Such 1952 transplants to Tuckerton were not reported to be successful in establishing breeding birds.

Modern records include birds at Cold Springs in August 1952, Brigantine Refuge* in April 1960 and again in April 1972, Stone Harbor in August 1966, and the Trenton Marshes* on April 13, 1976. Two were seen at Brigantine Refuge* May 13–24, 1975, and 2 at Manahawkin Wildlife Management Area* in June of 1976. A high count of 5 was obtained at Brigantine Refuge* on May 12, 1981, and at least 10 were in the state that month.

AMERICAN AVOCET *Recurvirostra americana* Gmelin

The Avocet is primarily a species of the western half of the United States. Formerly there was also a small coastal population of nesting summer residents in southern New Jersey. In the 1970s egg transplants (from Utah) were put into Willet nests, as with the Stilt experiment. Breeding birds did not become established, but it remains a future possibility. Now the Avocet is simply a rare migrant, seen most frequently in the fall (September–October).

The species was regular in our state from 1812 to 1869 although nesting ceased about 1829. There were fewer than 10 records between 1870 and the turn of the century, and the first significant modern record was of 12 at Newark Airport in September 1932. It has been recorded annually in recent years, even in small flocks at some locations (e.g. Brigantine National Wildlife Refuge* since 1960). Most records are still in the autumn, but summering birds were at Brigantine Refuge* in both 1959 and 1960 (6 birds each year).

A high count came from the fall of 1954 with 40 near Fortescue. This record is unlikely to be topped unless New Jersey again attains a breeding population.

SHOREBIRDS / *Scolopacidae*

GREATER YELLOWLEGS *Tringa melanoleuca* (Gmelin)

The Greater Yellowlegs is a common coastal migrant, but it also occurs regularly inland at ponds and lakes. Migration peaks mid-April to mid-May and in September. Small numbers of nonbreeding individuals remain through the summer. In recent decades small numbers have also been recorded in winter each year, almost entirely from the coastal areas.

A high count of 250 was recorded at Manasquan on May 28, 1927.

LESSER YELLOWLEGS *Tringa flavipes* (Gmelin)

Like the Greater Yellowlegs, this species breeds far to our north, but it is a common bird along New Jersey's coast during migration. Peak movement occurs April to May and August to September, and it is most abundant with the fall flights. Rarely, small numbers are noted in mid-summer or winter. (It is distinctly rarer than the Greater Yellowlegs on New Jersey Christmas counts.)

High counts include 335 at Hackensack Meadows* on August 3, 1975, about 600 at Newark Meadows* on September 18, 1929, and more than 700 at the Hackensack Meadows* in recent years, usually in late August.

SPOTTED REDSHANK *Tringa erythropus* (Pallas)

This accidental visitor from Europe has been recorded in only a few eastern states. There also have been identification problems with some reports (e.g. suggestions that "oiled" yellowlegs can look like a redshank).

New Jersey records are mostly from the Brigantine National Wildlife Refuge*: May–June 1965 (the first state record), late September 1978, and September 28–October 8, 1979. It is quite possible that a single bird was involved in the 1978 and 1979 sightings. There is also one record from Manahawkin, August 17–22, 1972. (And there is a nearby Pennsylvania record from Philadelphia at the Tinicum Refuge, July 1960.)

SOLITARY SANDPIPER *Tringa solitaria* Wilson

This shorebird is curious in many respects: it nests in boreal forests using other birds' nests, it is not social like most sandpipers, and it is more frequent inland than at the coast. In New Jersey it is strictly a migrant, particularly at inland ponds, streams, and lakes. It will pause for several days to a week at even the smallest bodies of water. It even appears frequently at stagnant ponds or ditches.

High counts of 13 came from cow sewage ponds at Plainsboro on May 10, 1968 and 40 at Black River on May 7, 1983.

WILLET *Catoptrophorus semipalmatus* (Gmelin)

The Willet is a migrant and summer resident along the coastal salt marshes. Migration peaks in late April to mid-May and in August. Usually it is more common in the fall migration.

Recent breeding populations began in 1952 after a 63-year absence of nesting in New Jersey. Today the species seems to be increasing in the Northeast, and it is found in loose colonies north to Nova Scotia. In our state Willet nests are usually on salt-marsh flats. (A study of New Jersey nesting preferences was published by Burger & Shishler in 1978.)

The western subspecies is distinctly paler and slightly larger than our eastern race. Western birds appear here in the fall mixed within flocks of eastern birds.

A high count of 1,000 Willets came from the Holgate Unit* of the Brigantine National Wildlife Refuge* on July 22, 1979.

SPOTTED SANDPIPER *Actitus macularia* (Linnaeus)

The Spotted Sandpiper is a common migrant at rivers, ponds, and lakes throughout (migration peaks in May and August). It is also widely distributed as a breeding bird, with grass-lined nests in pastures and fields, or at pond edges. (The females are often polyandrous.)

The species winters in southeastern states, but very rarely individuals are found in New Jersey (e.g. December 15, 1979, at Brigantine Refuge*).

High counts include 32 at Overpeck Creek,* Bergen County, on May 18, 1919, and 44 at Sandy Hook on July 21, 1976.

UPLAND SANDPIPER *Bartramia longicauda* (Bechstein)

This sleek sandpiper is a rare migrant and a rare local breeder in grassy pastures. Migration peaks in late April and in mid-July to mid-September. In the nineteenth century the species was much more common, and was widespread in agricultural fields. Formerly, for example, there were many nesting areas in central New Jersey. A 1981 survey of the state showed only 8–11 pair, in five localities, primarily in Salem and Somerset counties (Wander 1982).

High counts include 55 at Newark Meadows* in August 1934, 44 at Columbus on August 21, 1971, and 75 at Columbus on August 17, 1980. In recent years sod farms have provided migration staging areas, resulting in the high August counts.

ESKIMO CURLEW *Numenius borealis* (Forster)

This species is now extinct or on the verge of extinction. Presumably it was once an occasional fall migrant along our coast. During autumn migration it occurred regularly off eastern Long Island, New

York (Bull 1974). A New Jersey specimen was taken at Manasquan in the summer of 1880 (curiously, the specimen is deposited at the Denver Museum of Natural History). Although there have been several modern records (e.g. on coastal Texas) there is only the remotest possibility of any future sightings from our region. The last New Jersey reports were both from Cape May: one at the Coast Guard station on September 20, 1959 (details published in *Auk*, 1965, 82:493–496), and one at South Cape May on September 30, 1960 (listed in *Cassinia*, 1981, 59:47).

WHIMBREL *Numenius phaeopus* (Linnaeus)

The Whimbrel nests on the tundra of the high arctic and passes through our state as a coastal migrant. It is most frequent at salt marshes, with peak flights in May and mid-July to mid-August.

Birds of the white-rumped European race are accidental visitors to the Atlantic Coast states. Individuals have been irregularly reported in New Jersey during the past decade, and at least 4 were noted here in 1973.

A high count of 700 Whimbrels was made at Floodgates,* along the Delaware River, on August 25, 1978.

LONG-BILLED CURLEW *Numenius americanus* Bechstein

Although the Long-billed Curlew is a species of the western United States, in the nineteenth century it was a regular fall migrant on the Atlantic coast. It was last collected in New Jersey in 1898, and none has been seen after a few records in the early part of this century. (There was a recent sighting on Long Island, New York, during July 1975.)

BLACK-TAILED GODWIT *Limosa limosa* (Linnaeus)

This European species has occurred only a few times along the Atlantic coast of the United States. The New Jersey record, the fourth in North America, was well confirmed by many observers. The bird was seen at Brigantine National Wildlife Refuge* between May 22 and September 13, 1971 on many dates and by many observers. (During August all four godwit species were at the refuge.) Areas frequented by all of the godwits include the salt-marsh flats and freshwater pools.

HUDSONIAN GODWIT *Limosa haemastica* (Linnaeus)

As a breeder of the high arctic, this godwit is but a rare fall migrant along our coast. It is very rarely recorded in spring as it usually travels northward through central North America. It was quite rare at the turn of the century, and there were no New Jersey records from 1902 to 1909. There was a modest recovery and by the 1950s a few were seen each autumn.

High counts include 22 at Sandy Hook on September 6, 1979 (after Hurricane David) and 25 at Brigantine Refuge* on July 30, 1972.

BAR-TAILED GODWIT *Limosa lapponica* (Linnaeus)

This European species occurs as an accidental visitor on our coast. In the following list of records a single bird may have been seen at several locations or on subsequent dates. Early New Jersey records include birds at Beach Haven Inlet* in August 1951 and Brigantine Refuge* in the summer of 1937. Records from the 1970s include Somers Point (May 1972), Brigantine Refuge* (summer 1971, and June 19–30, 1974), and Longport (various years including May 6–20, 1973, May 14–19, 1975, April 28, 1977, and May 20–22, 1982).

MARBLED GODWIT *Limosa fedoa* (Linnaeus)

The Marbled Godwit breeds in pasture lands of the West, and it is only a rare visitor to northeastern states. During the nineteenth century it was more regular as a coastal migrant in New Jersey. However it declined and was very rare from 1890 to 1930. It increased slightly to become regular, but still rare, by the 1950s.

It is most frequently observed at coastal pools during the August–September fall migration. Some sightings have been made of a few winter birds, but it normally does not stay north of Virginia during mid-winter. Rarely, spring sightings are made, usually in May.

During its populational low it was often unobserved in New Jersey (e.g. there were no state records from 1921 to 1927).

RUDDY TURNSTONE *Arenaria interpres* (Linnaeus)

The Turnstone is a coastal bird at both its arctic breeding grounds and its winter range (south of New York). In New Jersey it is a common coastal migrant with peak numbers in May and in late August to September. It can be found in a wide variety of shore habitats, including open beach, rocky jetties, and mud flats. Small numbers occur in mid-summer and in winter (in southern New Jersey).

High counts have come from airplane surveys of the Delaware Bay shore where congregations of up to almost 100,000 occur during migration in late May. Like other shorebirds at this location Turnstones feed heavily on the abundance of freshly laid horeshoe-crab eggs.

RED KNOT *Calidris canutus* (Linnaeus)

The Red Knot is a shore migrant with peaks in May and in mid-August to mid-September. It occurs along the coast, but especially on the Delaware Bay shore. It is rarely found in winter, but sometimes small flocks occur near southern salt marshes.

A high count estimate comes from the Cape May bay shore with a remarkable 62,000 on May 27, 1981 (air survey). Our New Jersey migration flocks may involve a significant portion of the total North American population of Red Knots. They arrive along the Delaware Bay shore just in time to feed on an abundance of eggs from spawning horseshoe crabs.

SANDERLING *Calidris alba* (Pallas)

This cosmopolitan beach bird is a common migrant and a winter resident. Small numbers of nonbreeding individuals are often present in summer. Migration peaks in April and in August to mid-September, with the largest flocks on the outer beaches. Although Sanderlings are strongly associated with the ocean beach, they also occur on tidal flats or even freshwater areas. A spectacular high count of 11,760 was made on the Delaware Bay shore of New Jersey on May 27, 1982.

SEMIPALMATED SANDPIPER *Calidris pusilla* (Linnaeus)

This little sandpiper is a very common or abundant migrant on coastal mud flats and beaches, and to a lesser extent at inland bodies of water. Migration peaks in May and in late July through September. Small numbers are also seen in mid-summer, and even during the winter (in southern New Jersey). There has been considerable identification controversy over winter records which might involve the very similar Western Sandpiper; specimens are needed.

High counts include two large concentrations of "peep" (mostly Semipalmated Sandpipers) at Hackensack Meadows* on August 10, 1972, and July 31, 1975, both of which had thousands. An estimated 10,000 were noted at North Arlington on July 29, 1981, and an airplane survey of miles along the Delaware Bay shore estimated 114,000 on May 27, 1981.

WESTERN SANDPIPER *Calidris mauri* (Cabanis)

From its nesting grounds in western Alaska, the Western Sandpiper moves south to a winter range along both the Atlantic and Pacific coast states. In New Jersey it is a fall migrant (peak August–September) at mud flats along the coast, usually in flocks with other small shorebirds. It is generally quite uncommon as a migrant in spring. Small numbers are reported in mid-summer and winter from southern areas of our coast.

A high count of 1,000 came from Brigantine Refuge* on August 2, 1972.

RUFOUS-NECKED STINT *Calidris ruficollis* (Pallas)

Hypothetical. This Asiatic species may be an accidental visitor. There is a single sight record by three observers of a bird on tidal flats south of Beach Haven on July 15, 1962 (details were published in *Cassinia,* 1966, 49:30–31). Identification of this exotic species, which is so similar to the Western Sandpiper, requires much field experience and caution. Individuals occasionally stray southward from Alaska, but it is very rare even along the Pacific coast where it is more likely.

LITTLE STINT *Calidris minuta* (Leisler)

Hypothetical. This Eurasian shorebird has occurred as an accidental visitor to Alaska and to Atlantic Coast states (Delaware and Massachusetts). There have been two sight reports (most recently at Brigantine Refuge* on September 10, 1983), but more solid confirmation is needed before we can formally add this difficult to identify species to the state's list.

LEAST SANDPIPER *Calidris minutilla* (Vieillot)

This tiny shorebird is a common migrant on mud flats along the coast; it is also found in small numbers at inland ponds and lakes. Migration peaks in May and in late July through September. A few individuals are reported during winter months.

More than 1,000 have been estimated to be in large "peep" flocks during July and August.

WHITE-RUMPED SANDPIPER *Calidris fuscicollis* (Vieillot)

This small sandpiper is another long-distance migrant, from arctic breeding grounds to a winter range in southern South America. In its New Jersey passage it is an uncommon migrant at salt-marsh flats or freshwater puddles. Migration peaks in May and especially in mid-August through early October. Even then it is only found as individuals or in small flocks, usually within groups of other small sandpiper species.

A high count of 400 was obtained at Newark Meadows* on October 6, 1929. A summer high count of 90 was made at Thompsons Beach on June 9, 1982.

BAIRD'S SANDPIPER *Calidris bairdii* (Coues)

The Baird's Sandpiper is another species that migrates the length of the western hemisphere, but it is only with the autumnal flight that it passes our region (spring migration passes through the Midwest). As a rare fall migrant in New Jersey it is found mostly in grassy habitats with shallow pools (either fresh or salt water). Records are primarily from August through October.

A high count of 5 came from Cape May where the birds remained for a week in late August 1981.

PECTORAL SANDPIPER *Calidris melanotos* (Vieillot)

The Pectoral Sandpiper passes through New Jersey twice a year on its migrations between the high arctic of Canada and its South American wintering grounds. As a New Jersey migrant it occurs both at freshwater and saltwater pools, especially those with grassy edges. Peak numbers occur in April and May, and more notably from August into October. It has been seen into December, and even January (Cumberland County, 1981).

High counts include 100 at Brigantine National Wildlife Refuge* on July 20, 1980, 250 at Pedricktown on April 12, 1980, and 350 at Pedricktown on April 11, 1981.

SHARP-TAILED SANDPIPER *Calidris acuminata* (Horsfield)

This Asiatic species is an accidental visitor to Atlantic Coast states. In recent years there have been sightings at scattered localities from Massachusetts to Florida. There are three New Jersey reports, each of which involves two observers. Two are from Tuckerton (October 21, 1975, and August 6, 1981), and one is from Avalon Causeway* (September 21, 1980). It should be noted that this species strongly resembles the Pectoral Sandpiper, and variations in adult and juvenile plumages call for careful identification.

PURPLE SANDPIPER *Calidris maritima* (Brünnich)

The Purple Sandpiper is remarkably adapted to cold coasts: it breeds on the shores of the Arctic Ocean and some even remain through the winter in Greenland. In our area it is a local winter resident on rock jetties along the coast. (Very rarely it is found on mud flats of the Delaware Bay shore, up to Gloucester County.) It can be present from October to May, but it is most frequently encountered in the winter months of December to February.

High counts for the Purple Sandpiper include 110 at Brigantine Island* on February 1, 1976, and 120 at Cape May on May 3, 1964.

DUNLIN *Calidris alpina* (Linnaeus)

The Dunlin comes south from arctic breeding grounds to winter along the Atlantic and Gulf coast states. In our area it is a common migrant and winter resident along the coast. Migration peaks in May and in October–November, often with large flocks. A few nonbreeding birds remain in summer. It is the most common shorebird in winter.

High counts include estimates of 10,000 on the Cape May Christmas count in 1966 and again at Stone Harbor in February 1971. A peak migration census was made along the Delaware Bay shore with a remarkable 13,500 on May 20, 1981 (airplane survey).

CURLEW SANDPIPER

Calidris ferruginea (Pontoppidan)

This European species is an accidental migrant, but it is now recorded almost annually on the New Jersey coast, usually with several sightings. Most reports are in the spring (e.g. May at Tuckerton), but records also come from summer and fall (e.g. Barnegat Light on October 28, 1934). There are a number of historical reports as well: Audubon listed 2 shot at Great Egg Harbor* in the spring of 1829, and other state specimens were taken in the mid-nineteenth century.

High counts include 3 at Brigantine Refuge* (May 14–22, 1963), 3 at Hereford Inlet* (September 6, 1966), and several in southern New Jersey (spring of 1977). There were also 3 reported from coastal New Jersey in May 1981 and again in May 1983.

STILT SANDPIPER *Calidris himantopus* (Bonaparte)

This uncommon shorebird is a regular fall migrant (peak movements August through September) along the coast. It frequents shallow pools with other shorebirds, especially yellowlegs. Unfortunately we seldom see its fine breeding plumage since it is decidedly rare in our area during spring migration (most spring records are from May).

A high count of 600 was reported from Brigantine Refuge* in late July 1981.

BUFF-BREASTED SANDPIPER

Tryngites subruficollis (Vieillot)

The Buff-breasted Sandpiper has a very restricted breeding range in tundra areas along the Arctic Ocean. It is scarce even within its main migratory route along the Midwest, and in New Jersey it is a very rare fall migrant (late August–September). Like the Upland Sandpiper, it favors short-grass areas such as sod farms during migration. It is a tame species and close approach is often possible.

A remarkable high count of 17 was made at Cape May in early August 1981.

RUFF (MALE) and REEVE (FEMALE)

Philomachus pugnax (Linnaeus)

This species is a very rare migrant, coming to the Atlantic Coast area from Europe. (There might also be a small breeding population in northern Canada.) Most reports are from April and August. There are also a small number of mid-summer sightings.

The first New Jersey record is from Barnegat in 1869 (specimen). Early sightings include Tuckerton (2 on October 2, 1932), Beach Haven (July 2, 1944), Bridgeport (May 3–8, 1947), and Elizabeth (October 3–24, 1948). The species became "regular" in the 1950s, and there were several being recorded annually by the 1970s. High counts are facilitated by the distinctive plumage variations that often allow individual recognition. Such tallies include 3 at Pedricktown (March 26–April 3, 1979), 4 at Mannington Marshes,* Salem County (April 1968), and 5 (4 Ruffs) near Pedricktown (April 12, 1981).

SHORT-BILLED DOWITCHER

Limnodromus griseus (Gmelin)

The Short-billed Dowitcher is an abundant migrant, with peak numbers in May and in mid-July through August. Smaller numbers can be found in June, when there is even some overlap of late northbound migrants with early returning birds! Some Short-billed Dowitchers are present through September and October, and a few are usually tallied on the state's Christmas counts. (See Long-billed Dowitcher for late fall reports.)

High counts are made of large flocks at tidal mud flats: 25,000 were tallied at Brigantine Refuge* on August 2, 1972.

LONG-BILLED DOWITCHER

Limnodromus scolopaceus (Say)

The Long-billed Dowitcher breeds in the very northwestern corner of North America, along coastal Alaska. Some pass our area as uncommon coastal migrants, usually later in the fall (October–November) than the Short-billed Dowitcher. It is very rarely reported in New Jersey during spring migration, but there are scattered records from May. (The great similarity between the two dowitcher species necessitates careful identification, usually by call notes.)

A high count of more than 100 was made at Brigantine Refuge* in October 1981.

COMMON SNIPE *Gallinago gallinago* (Linnaeus)

The Common Snipe is a migrant in our wet habitats such as marshes and flooded meadows. Migration peaks in March–April and October–November. Small numbers are also recorded in winter where there is flowing water (e.g. at marsh springs). New Jersey is just at the southern edge of the Snipe's breeding range, and its nesting here is now limited to a few wet-field areas in northwestern counties. (The first modern breeding record came from Sussex County in 1950.)

A maximum of 500 was estimated at the Great Swamp* on March 30, 1972, and a winter high of 69 comes from the Trenton Marshes* Christmas count on January 3, 1981.

GREAT SNIPE *Gallinago media* Linnaeus

Hypothetical. This European species has been reported as an accidental visitor, but the single New Jersey record has been questioned. A photograph of a snipe thought to be this species at Cape May on September 7, 1963, may not be conclusive. Details were published in *Cassinia* (1966, 49:20–23), but opinions of experts varied. The only other Atlantic Coast record, from Virginia, has been dismissed as insufficient (*Auk,* 1981, 98:643).

EURASIAN WOODCOCK *Scolopax rusticola* Linnaeus

This large woodcock is an accidental visitor from Europe. Most North American records are from the nineteenth century (including New Jersey, Pennsylvania, and Virginia). Two of our old state records are from game dealers (e.g. near Shrewsbury on December 6, 1859). In modern times there was one at Goshen, Cape May County, from January 2 to 5, 1956. (The only other North American record from this century is an Ohio sighting in 1935.)

AMERICAN WOODCOCK *Scolopax minor* Gmelin

The American Woodcock is a common migrant in woodlands and hedgerows throughout. It prefers moist habitats, and it uses open areas near woods for courtship display flights. Its migration peaks in early March and November. It is a local breeder in the northern half of the state, with active nests as early as mid-March. There are wintering populations near the coast in southern New Jersey. (In some years Cape May has had the nation's highest Christmas count of woodcocks.) This game bird apparently suffered from excessive hunting in the nineteenth century.

High counts include 350 at Cape May on November 4, 1970, and 2,000 at the Great Swamp* November 10–16, 1970.

WILSON'S PHALAROPE *Phalaropus tricolor* (Vieillot)

The Wilson's Phalarope nests at freshwater ponds and lakes in the midwestern prairies. It comes to our area as a rare migrant at coastal ponds, or inland at freshwater pools. It is mostly seen during the fall (August–October), but it is also regularly recorded in spring (May and early June). The Wilson's Phalarope is more terrestrial than the other phalarope species and will often forage on mud flats.

A high count of 6 was reported at the Newark Meadows* on September 30, 1934.

RED-NECKED or NORTHERN PHALAROPE *Phalaropus lobatus* (Linnaeus)

Like the Red Phalarope, this species breeds far to the north and migrates to little-known wintering areas at sea. As with all the phalaropes, the female of this species is the more brightly colored of the sexes. It is primarily an uncommon offshore migrant (May and July through September), at coastal ponds. It occurs more rarely at inland bodies of water, or along rivers (e.g. Raritan River,* New Brunswick on September 21, 1976).

RED PHALAROPE *Phalaropus fulicaria* (Linnaeus)

The Red Phalarope breeds in the high tundra of northern Canada and winters in southern latitudes at sea. In New Jersey it is an uncommon offshore migrant (April–early May and September–October). At infrequent intervals this pelagic species is brought to coastal land with strong northeast winds. Also, rarely, there are summer and winter sightings of small numbers at sea.

High counts all come from the far offshore pelagic zone (e.g. Hudson Canyon* where there was a spectacular maximum of 17,000 on April 18, 1980).

JAEGERS, GULLS, AND TERNS
Laridae

POMARINE JAEGER

Stercorarius pomarinus (Temminck)

From the tundra nesting areas this species comes south as a pelagic migrant, and it is usually seen at offshore fishing fleets from September to mid-November. All three jaeger species are predatory birds which are frequently seen chasing gulls or terns to rob them of fish. (This conspicuous behavior often helps observers first note the jaegers' presence within groups of gulls.) In addition to fall pelagic sightings, rare shore sightings also occur (e.g. at Island Beach State Park*). There are some records from spring migrations as well.

High counts are from the far offshore zone (e.g. 51 at the Hudson Canyon* on September 17, 1972).

PARASITIC JAEGER

Stercorarius parasiticus (Linnaeus)

The Parasitic Jaeger is primarily a pelagic migrant, uncommon or rare along the shore. This is, however, the jaeger species most frequently seen from the beach, often in pursuit of terns or gulls. Migration peaks in September–October and more rarely in May. Very rarely there are even inland records at lakes or reservoirs. (Jaegers have been seen along the Delaware as far north as Trenton.)

High counts include 50 passing Island Beach State Park* on September 20, 1968, and about 70 passing Point Pleasant (Ocean County) on September 19, 1926.

LONG-TAILED JAEGER

Stercorarius longicaudus Vieillot

This species, the smallest of the jaegers, is a rare pelagic migrant, usually in the far offshore zone from August to October. It is very rarely seen along the coast, in migration with other jaeger species. (In a season's tally of 50 migrating jaegers one would expect more than 40 Parasitic Jaegers but only 1 or 2 Long-tailed Jaegers.)

GREAT SKUA *Catharacta skua* Brünnich

This large pelagic visitor has an uncertain status off our coast. It breeds in the high arctic, and it is rare but possibly regular off New Jersey. Skuas are regularly observed at deep-sea fishing fleets, but it is usually not known which species is involved (see South Polar Skua). There are regional records of the Great Skua for most months of the year.

The first state record came from 5 miles off Avalon, June 13, 1969. High counts include 15 at fishing fleets about 80 miles offshore (January–February 1976), and 20 about 75 miles off Sea Isle City, mid-January 1977.

SOUTH POLAR SKUA

Catharacta maccormicki (Saunders)

Like the Great Skua, this is a pelagic species of uncertain status. It breeds in the Antarctic and comes northward to our area of the Atlantic as a rare but possibly regular transient. (This form was considered by some to be conspecific with the Great Skua.) Most skuas in deep waters off Long Island, New York, are now being identified as this species.

Records include 1 seen by many observers on a pelagic trip 70 miles east of Corson's Inlet* on May 27, 1978, and 2 well studied at the Hudson Canyon* on May 28, 1983.

LAUGHING GULL *Larus atricilla* Linnaeus

The Laughing Gull is a coastal summer resident, common as a breeding bird on southern New Jersey salt marshes. It is generally present from April through October. (The species has been expanding northward during recent years, and it is now nesting on Long Island, New York, and even in coastal Maine.) Laughing Gulls are rare in our area during winter, when most individuals are found south of the Carolinas. A remarkable movement of an immature Laughing Gull was recorded in 1980 when a banded bird from New Jersey was found five months later on Kauai, Hawaii!

High counts include 5,200 at a colony on a salt-marsh island at Brigantine National Wildlife Refuge,* in 1978. A 1979 survey of the state's coast showed 53,387 nesting birds in 67 colonies. There is now concern about the negative effects of expanding Herring Gulls on our Laughing Gulls, however (Burger 1979b), and colonies should be carefully watched in the future.

FRANKLIN'S GULL *Larus pipixcan* Wagler

The Franklin's Gull nests in midwestern prairies and winters on coasts far south of the United States. In our area it is an accidental visitor. The first record for New Jersey was an adult in breeding plumage at North Arlington, February 9–17, 1975. Subsequent sightings are mostly from Cape May, with many dates (e.g. much of August 1979, August 9–October 6, 1980, an immature on August 27, 1981, and an adult on September 24, 1981). There have been numerous records from western portions of Pennsylvania and New York that include every month.

LITTLE GULL *Larus minutus* Pallas

This tiny gull is a rare visitant along the coast. It is a European species but since 1962 small numbers have been discovered breeding in North America (e.g. at Lake Ontario and Lake Erie). Our regional records come from every month, but it is most likely from winter into early spring.

New Jersey's earlier reports include a May 1929 bird at Newark Meadows,* 1 at Point Pleasant on August 11, 1929, and 1 at Jersey City on August 22, 1937. Records in the 1950s come from Barnegat Inlet,* South Amboy, Jersey City, and Sandy Hook.

High counts range from 4 (e.g. South Amboy in May–June 1962, and Cape May in March–April 1980) to 12 (immatures) at Caven Cove* on May 21, 1980.

COMMON BLACK-HEADED GULL

Larus ridibundus Linnaeus

The Black-headed Gull is primarily a European species, although there is a small breeding population in North America (primarily in Newfoundland). In New Jersey it is a rare visitant, partial to coastal inlets and bays. Most records are from November to April, but summering individuals are sometimes seen, albeit rarely. The Black-headed Gull often frequents sewage outlets with other gull species.

The first New Jersey record is from near Laurence Harbor in Middlesex County on November 21, 1948. The species was seen almost annually thereafter along the central New Jersey coast. Records of particular interest include 1 at Cape May (January 2, 1956), 4 at Jersey City (March 9, 1958), 6 (5 immatures) at Shark River Inlet* (March 19, 1961), and 24 on the Hudson River (December 21, 1976). It can be found regularly at some localities now (e.g. near Liberty State Park*). Summer records have come from South Amboy in late June 1963 and August 1956, and from Cape May in August 1981.

BONAPARTE'S GULL

Larus philadelphia (Ord)

From nesting grounds in northwestern parts of the continent the Bonaparte's Gull comes to our area as a common migrant and winter visitant, mostly along the coast. Migration peaks in April and in November. Rarely, stragglers occur along the shore in summer. Inland lakes occasionally have small flocks during spring and fall migrations.

High counts include 1,000 at Edgewater on November 27, 1966, a Christmas count of 1,177 at the Raritan Estuary* on December 31, 1977, and 3,000 at Sandy Hook on December 28, 1975.

MEW GULL

Larus canus Linnaeus

The Mew Gull comes to our area as an accidental visitor from either the western United States or Europe. We have a small number of sight reports from our state, but further confirmation is desirable. The first report came from Jersey City on October 31, 1955. (Since the bird was a difficult to identify subadult some have considered this record insufficient.) Another immature (first-year bird) was seen at Seven Mile Beach* on October 5, 1974. Four were reported off Sandy Hook on December 18, 1977, and an adult was identified at the Cape May Canal* on October 10, 1980. (The Cape May bird was identified as the European subspecies.)

RING-BILLED GULL *Larus delawarensis* Ord

The Ring-billed Gull is a common migrant and winter visitant, but it is also numerous in summer (nonbreeding birds). It is primarily a coastal bird, but can also be frequently seen inland at lakes and plowed fields, especially during migration or after storms. (In comparison with the Herring Gull, the Ring-billed is relatively more common inland.) This species has increased dramatically over recent decades, and nesting colonies are found as close as the lakes of upstate New York.

Inland high counts include more than 400 at Lake Assunpink,* November 1980, and 650 at Round Valley Reservoir* on November 3, 1981, and 8000 near Camden on April 4, 1983.

CALIFORNIA GULL *Larus californicus* Lawrence

The California Gull is an accidental visitor from the western United States. New Jersey has had only a few reports: a bird seen by several observers at Brigantine National Wildlife Refuge* on August 13, 1967, and a possible adult of this species at Great Egg Harbor* on August 20, 1981. (There were sightings in New York State during 1978 and 1979.)

HERRING GULL *Larus argentatus* Pontoppidan

The Herring Gull is an abundant winter visitor, mostly along the coast, from September through April. It is also common as a nonbreeding resident in summer (mostly immatures). In recent decades nesting colonies have been established along the shore (Bongiorno & Swinebroad 1969; Burger 1977). The first New Jersey nesting came from Stone Harbor on July 14, 1946 (five eggs, possibly from 2 birds). More shore nestings were recorded in 1949, 1950, and 1954. It continued to become increasingly common as a breeding species, and a state survey in 1979 showed 5,839 nesting in 55 colonies. (Its increase is of some concern because it can be a significant predator on the chicks of terns and other gulls.)

High counts include 15,070 on the Raritan Estuary* Christmas count (December 31, 1977), and 25,000 at the far offshore fishing fleets (January–February 1976).

THAYER'S GULL *Larus thayeri* Brooks

The New Jersey status of the Thayer's Gull is uncertain, but it is probably a very rare winter visitor along the coast. This little known arctic species was formerly considered a race of either the Herring Gull or the Iceland Gull.

The first state record is of a bird taken at Mount Ephraim, Camden County, on March 9, 1888 (specimen in the Academy of Natural Sciences, Philadelphia). There have been almost no other published records, but recent recognition of this gull as a separate species has increased the awareness and interest of field observers. Recent sightings include 1 at Sylvan Lake,* Avon, on February 24, 1976, 1 photographed at Lily Lake,* Cape May in December 1980, and another at Cape May (first-year bird) on April 28, 1982.

ICELAND GULL *Larus glaucoides* Meyer

The Iceland Gull nests in Greenland and northern Canada. Its normal wintering area extends southward only to Maryland, and in New Jersey it is an uncommon winter visitor near the coast. Most sightings occur from December to April. The *kumlieni* race, with grayish primaries, predominates in our area.

High counts include 8 at coastal lakes near Asbury Park (January 1973), 13 on the Delaware River (February 9, 1963), and 9 at North Arlington (December 1975). The North Arlington count included 2 first-year birds, 4 second-year birds, 1 subadult, and 2 adults. (Generally, adult birds are quite scarce in our state.)

LESSER BLACK-BACKED GULL *Larus fuscus* Linnaeus

This European species is a very rare visitor to the coast in fall and winter (mostly from October to March). Many of the records are from garbage dumps, where it associates with other gull species.

The first United States record came from New Jersey: a bird at

Beach Haven on September 9, 1934. Subsequently there are records from Newark Bay* (February 7, 1937), Beach Haven Inlet* (December 2, 1950), North Arlington (2 in the winters of 1961 and 1981), Jersey City (December 1974), Belmar (December 1974), and Lincoln Park (January 2, 1972).

High counts include 3 at Johnson Park,* New Brunswick in the fall of 1979 and the early spring of 1982. A remarkably early high count of 3 was made at the Raritan Bay* in November 1950, and it is only in recent years that several new highs are being reported. During the fall of 1981 there were at least 4 Lesser Black-backed Gulls in southern New Jersey (including 1 at Cape May Point).

GLAUCOUS GULL *Larus hyperboreus* Gunnerus

The Glaucous Gull is one of our most northerly species. It breeds in the high arctic (e.g. coastal Greenland), and usually it comes only as far south as the mid-Atlantic states in winter. As a rare winter visitor it is most often observed near the coast. A review of New Jersey records shows that most involve pale immature birds, from December to March.

A maximum of 10 at the far offshore fishing fleets was reported in January–February 1976. High counts are also obtained at garbage dumps where it frequently feeds. The species is slowly increasing in frequency within our area.

GREAT BLACK-BACKED GULL *Larus marinus* Linnaeus

The Great Black-backed Gull is primarily a common winter visitant (September through April), but it has become increasingly numerous as a breeding summer resident in recent years. The first New Jersey nesting record came from Absecon Bay* in July 1966. It continued to increase slowly (Burger 1978b), and by 1979 there were at least 130 nesting along our coast. (The southward expansion of its breeding range has already reached North Carolina.)

An impressive high count came from the far offshore fishing fleets in January–February of 1976 with 2,000 birds. Inland, a remarkable 800 were tallied on Lake Parsippany,* Morris County, in mid-February 1981.

BLACK-LEGGED KITTIWAKE

Rissa tridactyla (Linnaeus)

This northern gull is a pelagic migrant and visitant, with the largest concentrations during the winter (November–January) in the far offshore zone. In our area it is rarely seen from land. The species is having a population expansion in the north, and small numbers of nonbreeding birds are now being seen off New Jersey during summer months.

High counts include 3,000 passing Island Beach* in three hours on November 13, 1972, 6,000 at offshore fishing fleets in December 1975, and about 15,000 75 miles off Sea Isle City, January 14–16, 1977.

SABINE'S GULL *Xema sabini* (Sabine)

The Sabine's Gull is an accidental or very rare pelagic visitant from the arctic. (The majority of Sabine's Gulls migrate to and winter in the Pacific.)

The first New Jersey record is the only one from land: a bird at Seaside Park on November 8, 1936. Subsequent records were obtained on boat trips 24 miles east of Cape May Inlet* in October 1979, and off Atlantic City in October 1979. (Records from New York State are primarily from mid-September through mid-October.)

IVORY GULL *Pagophila eburnea* Phipps

This circumpolar gull remains in icy waters the year round. It is only an accidental winter visitor as far south as New Jersey. We have two notable records for our state: 1 found dead at Island Beach* on February 3, 1940, and 1 seen by several observers at Manasquan Inlet* on January 30, 1955. One was also recently reported but unverified, on a January 2, 1982, Christmas count far offshore (east of Belmar).

GULL-BILLED TERN *Sterna nilotica* Gmelin

The Gull-billed Tern is widely distributed throughout the world, but breeding colonies are often very local and sporadic. It is a rare summer resident along coastal salt marshes in southern New Jersey, May through August. It has been a very rare breeder in recent years.

In 1811 it was reported breeding in some numbers at Cape May County. After about 1870 it was quite rare in the state, and there were almost no records until 1926 when a nest with two eggs was found at Stone Harbor in June. Again, there were almost no reports until there was suspected breeding at Brigantine National Wildlife Refuge* in 1956. Stone Harbor had one nesting pair in 1958 and 1959. A coastal breeding bird survey in 1979 still showed only a few of this species. (The first New York State nesting occurred on Long Island in 1975.)

Coastal summer congregations have yielded high counts, such as 20 at Brigantine Refuge* in 1976. Inland, at cranberry bogs in the Pine Barrens,* there have been remarkable concentrations at the end of recent summers. For example, at Whitesbog (25 miles from the coast) there were 30 in August 1980 and 70 on August 10, 1981. At the cranberry bogs the terns characteristically forage by hawking insects.

CASPIAN TERN *Sterna caspia* Pallas

This cosmopolitan tern is an uncommon migrant in fall (peak from mid-August through September), and a rare migrant in spring (May). The nearest nesting areas are at the Great Lakes, and very locally on Virginia beaches. The vast majority of the sightings are coastal, but it has been recorded at scattered localities inland (e.g. Princeton, Assunpink*) and along the Delaware River (e.g. Bull's Island State Park*). Rarely, individuals are seen in mid-summer, or into December.

A high count of 13 was reported from Seven Mile Beach* in 1933, and recently a count of 21 was reported along the Cape May coast on September 4, 1981.

ROYAL TERN *Sterna maxima* Boddaert

This colonial tern of southern shores is a coastal summer visitor and fall migrant (peak movement late August through early October). Formerly it was very rare as a New Jersey visitant, usually after hurricanes. It has become quite regular and might well become a rare nesting species in the future. It currently nests as nearby as

Maryland, but published reports of New Jersey nesting in 1975 are apparently incorrect.

In 1869 the Royal Tern was described as very rare in our region, and records were associated with hurricane passage. Even in the first half of this century sightings were scarce and usually storm associated (e.g. a flock at Brigantine Beach* on September 17, 1933). Since 1950 there has been a gradual but continuous increase in sightings.

Inland reports are quite rare (e.g. Warren Grove on September 17, 1944, after a hurricane). Similarly, spring reports have been few until recent years.

High counts include more than 200 at Stone Harbor on September 7, 1979 (associated with a hurricane), and more than 500 at Cape May on October 7, 1979.

SANDWICH TERN *Sterna sandvicensis* Latham

The Sandwich Tern regularly breeds as far north as Virginia, but it is only a very rare tropical visitant to the southern New Jersey coast. The species formerly bred in Maryland, and recent summering birds in New Jersey could conceivably be nesting in the future. Sightings are increasing in frequency, and there are now records every year along our coast.

The first New Jersey record came from Five Mile Beach,* in August 1861. More recently records have come from Beach Haven Inlet* on September 1, 1954 (Hurricane Carol), Tuckerton on September 26, 1968, Stone Harbor on September 27, 1964, and North Cape May on an unusual spring date of April 16, 1972. Birds have been present during part of the summer at Holgate (2 including 1 immature in 1977, and 1 in 1979), and at Cape May (1979).

ROSEATE TERN *Sterna dougallii* Montagu

The Roseate Tern is found in the Atlantic, Pacific, and Indian oceans, but it is a very local breeding bird. (The world's largest breeding colony is probably that of the northern U.S. Virgin Islands.) The species is a very rare summer transient along the New Jersey coast and is extremely rare as a breeder. It is most likely to be seen during spring migration (May into June) along the coast. It is unclear if the Roseate Tern was ever more common as a nesting species in southern New Jersey (e.g. during the 1880s) because the old records based on egg identification might not be reliable. (In the northeastern United States it nests along the coast of New England south to Long Island, New York.)

Each summer there are few if any pair along our coast (e.g. at Holgate). In 1981 3 were seen in Cape May during early July, but nesting was not suspected. A high count of 40 was made in an early September migratory flock of 1,000 Common Terns at Island Beach.*

COMMON TERN *Sterna hirundo* Linnaeus

This coastal tern is a common summer resident along beaches from mid-April through September. Very rarely individuals may remain into winter. Inland reports at lakes are infrequent (usually July–September). Hurricane-driven birds may even appear inland far away from any water (e.g. Kendall Park).

The 1979 coastal breeding-bird survey tallied 9,581 Common Terns in 90 colonies. Colony size ranges from a few pair to 300, but most are about 50 pair. Often the Common Tern nests in close proximity to other beach nesting birds such as the Black Skimmer (e.g. at Holgate). There has been concern about declines in the Common Tern caused by predation and competition from the increasing numbers of breeding Herring Gulls along the New Jersey coast.

ARCTIC TERN *Sterna paradisaea* Pontoppidan

The Arctic Tern is a very rare pelagic migrant. Its status is uncertain because of its far offshore migration and difficulty in species identification (especially in the fall). Very rarely it appears at beaches within flocks of Common Terns.

Records of varying levels of confirmation include Absecon Inlet* (October 25 and November 12, 1975), Sandy Hook (May 25, 1975), Stone Harbor (August 2, 1973, and September 1980), Tuckerton (July 19, 1961), Barnegat Light (November 27, 1981), and Cape May (August 1973, September 1980, and June 1982). Good photographic records were obtained at South Amboy on May 30, 1970, and at Cape May in June 1982.

Offshore records include sightings from the Hudson Canyon* (June 5, 1976, and May 28, 1977), and 24 miles off Cape May on October 14, 1979. In May 1983 2 were seen on a Hudson Canyon* trip.

FORSTER'S TERN *Sterna forsteri* Nuttall

The Forster's Tern has a very limited breeding range along the Atlantic coast, from southern New Jersey to Cape Charles, Virginia. In our state it is a summer resident along the coastal salt marshes. Often it is common into the fall (mid-August to mid-October) and rarely into December. Recently it has increased as a nesting species. (It bred on the South Jersey coast in the last century.) The first modern breeding came from several pair at Brigantine Refuge* in 1955–1956. By 1979 there were about 1,167 in 18 coastal colonies.

A high colony count of 200 was made at a remote section of Brigantine Refuge* in the summer of 1978. A remarkable post-breeding concentration of 700 occurred on the bay shore of North Cape May on July 14, 1981.

LEAST TERN *Sterna antillarum* (Lesson)

The Least Tern is a coastal summer resident, nesting on open beach areas. It is primarily limited to the period from May through September, and it winters in the South Atlantic (off South America).

The Least Tern is sensitive to disturbances, and nesting colonies are often precarious. It has declined throughout much of its range, and it is frequently listed as an endangered breeding species. It is reported to have bred commonly in New Jersey through much of the nineteenth century. In the 1880s, however, Least Terns were killed by plume hunters for the millinery trade, and no nesting reports came from our state for three decades (1890–1919). Improvements began in the 1920s, and by 1924 there was a colony of about 20 pair in Cape May County. By 1925 there were 5 small colonies in southern New Jersey. By 1960 the Stone Harbor colony had increased considerably, and in 1971 the Hackensack Meadows* reported 80 birds. By 1979 there were 1,037 in a coastal survey, nesting in 14 colonies. The next year's survey estimated 2,169 breeding in the state. In 1981 20 colonies held 1,600 adults, with a colony maximum of 350 at Corson's Inlet,* Ocean City.

Taxonomic note: Some field guides refer to this as the Little Tern, for at one time the Least Tern was considered conspecific with its European relative, *Sterna albifrons.*

BRIDLED TERN *Sterna anaethetus* Scopoli

Like the Sooty Tern, this is a very rare coastal visitant from tropical seas, and its appearance in this region is highly correlated with hurricane passage.

Reports include a February 1951 carcass at Island Beach,* 1 at Lavalette on August 13, 1955 with 8 Sooty Terns (Hurricane Connie), Manasquan Inlet* on September 11, 1954, 5 at Island Beach* in September 1960 (Hurricane Donna), 4 at Brigantine National Wildlife Refuge* on September 4, 1966, 1 offshore from Barnegat Light on August 15, 1976 (Hurricane Belle), 2 off Barnegat Inlet* on August 14, 1979, 1 at South Cape May on September 7, 1979, and 1 dead near Atlantic City on October 8, 1979. (The last two records were related to Hurricane David.)

SOOTY TERN *Sterna fuscata* Linnaeus

The Sooty Tern is a very rare coastal visitant from tropical seas. Its appearance in the Northeast, like that of the Bridled Tern, is usually associated with hurricanes.

Records include Long Beach in the 1870s, Corson's Inlet* on September 7, 1916, Seven Mile Beach* on January 13, 1929 (decomposed specimen), Stone Harbor on September 23, 1980 (immature), and Cape May on September 28, 1980 (adult).

Hurricane-associated records include a dozen between Sandy Hook and Cape May (Hurricane David, September 1979), a dozen records along the coast with Hurricane Connie in August 1955, and 35 at Island Beach State Park* from Hurricane Donna in 1960. Single records come from Belmar on August 15, 1976 (Hurricane Belle), and from Cape May in June 1972 (Hurricane Agnes).

WHITE-WINGED TERN *Chlidonias leucopterus* (Temminck)

Hypothetical? This European species has been recorded as an accidental on the Atlantic coast (e.g. Virginia and Delaware) in recent summers, and it should be expected in New Jersey. We have older unconfirmed reports from our coast (Raritan Bay, Manahawkin, and further south), and more recently at Cape May and Sandy Hook, both in May 1983.

BLACK TERN *Chlidonias niger* (Linnaeus)

Unlike other eastern terns, this species nests in freshwater marshes, as nearby as upstate New York. Most individuals have a migration route through the Mississippi Valley rather than along the coast, so the Black Tern is only a scarce migrant in New Jersey. It is generally rare in spring migration (May), and uncommon in fall (July through September), at both inland lakes and marshes, as well as the coast. It is occasionally common offshore during fall migration. Since its normal wintering range is in the southern Caribbean the Black Tern is only accidentally recorded in New Jersey during the winter (e.g. 1 on a 1965 Barnegat Christmas count).

BROWN NODDY *Anous stolidus* (Linnaeus)

The Brown Noddy is an accidental visitor from tropical seas. There are but a few records of this dark tern from New Jersey. Most state records were associated with Hurricane Donna in 1960, when many observers saw this species at Stone Harbor and Island Beach.* More recently there was a report by two observers from Cape May on September 8, 1979.

BLACK SKIMMER *Rynchops niger* Linnaeus

This species is strictly coastal, and it is a summer resident along our beaches and bays (May through October). It nests in colonies on the barrier beaches with four to five eggs laid in a sandy depression.

In the nineteenth century it was a common breeder, but it was very rare by 1890 because eggs were stolen for food. In southern New Jersey it went unrecorded from 1895 to 1915. In the decade of 1927–1936 it increased markedly, and expansion continued through the 1960s and 1970s. Very rarely there are winter records (e.g. January 30, 1980, at Holgate).

Recent (1979 and 1980) coastal surveys have shown about 1,800 nesting in 15 colonies in New Jersey. But the 1981 state census had only 1,300 adults, which produced 500 young.

A high count of 2,000 came from the Avalon Causeway* colony in the summer of 1970. The Holgate colony had 600 individuals in the summer of 1977 and again in 1983.

AUKS / *Alcidae*

DOVEKIE *Alle alle* (Linnaeus)

The Dovekie nests in the high arctic (e.g. coastal Greenland), and it winters at sea south to New Jersey. Usually this tiny auk is a rare winter visitant along our coast, but it might be found regularly offshore, since it has pelagic habits. Infrequently, there are flight years when the species comes south in greater numbers (most records from December to February). Two great invasions were in the winter of 1932 and in December 1966. In the latter about 40 were recorded along the central coast, and probably more than 100 were in the state. Storms can cause inland records at lakes or reservoirs, but most recent sightings are strictly maritime.

COMMON MURRE *Uria aalge* (Pontoppidan)

The Common Murre is a seabird of cold waters in the North Atlantic and North Pacific oceans. Almost everywhere it is less common than the Thick-billed Murre, and in our region it is extremely rare as a winter visitant along the coast and offshore. Records include Avon on February 3, 1946, Shark River Inlet* on December 8, 1951, Tuckerton on February 15, 1959 (dead), and two at coastal Cape May County on December 24, 1967. Except for New York, our neighboring states also have very few records for this species.

THICK-BILLED MURRE *Uria lomvia* (Linnaeus)

The Thick-billed Murre is the more northerly nesting of the two murre species, and it winters a bit further south (regularly to New Jersey). Still, it is only a very rare winter visitor along the coast and at sea. Each year we have few if any records within the state. Most sightings are from December (e.g. Cape May Point, December 26, 1976), but there are also reports into the late spring (e.g. coastal Cape May County, May 11, 1952). A remarkable high count of 10 was obtained 10 miles off Sea Bright on December 27, 1908.

RAZORBILL *Alca torda* Linnaeus

The Razorbill nests on sea cliffs of Canada and Greenland, and it winters at sea south to our shores. Still, it is only a rare winter visitor (December to March) along our coast and offshore.

Reports include Christmas counts (e.g. Cape May in 1959), oiled birds on beaches (e.g. 6 on Long Beach Island, Ocean County, in the spring of 1973), and offshore (e.g. 5 at the Hudson Canyon* on March 5, 1977). A high count of 40 was recorded on a pelagic trip out from Cape May in March 1981.

GREAT AUK *Pinguinus impennis* (Linnaeus)

This large coastal alcid of the North Atlantic is extinct. It formerly moved south in winter to northeastern states (regularly to Massachusetts and possibly to New Jersey). Although we have no definite record it seems probable that it occurred, at least accidentally, along our coast. Human persecution brought the species to extinction about 1844.

BLACK GUILLEMOT *Cepphus grylle* (Linnaeus)

This Guillemot nests as nearby as the rocky islands off Maine, but it has very little southern movement after breeding. Therefore it is only a very rare winter visitor along our coast. New Jersey records include Spring Lake, Monmouth County (December 27, 1958), Newark Bay* (December 27, 1936), Boonton Reservoir* (November 26, 1939), and 2 at Manasquan Inlet* (April 23, 1977). (Sightings are often at rocky jetties rather than open beach.)

ATLANTIC PUFFIN *Fratercula arctica* (Linnaeus)

This alcid breeds commonly in areas of coastal Labrador and Newfoundland. Smaller colonies nest as far south as Nova Scotia and even Maine. (Reintroduction programs have been important for U.S. breeding populations.) In winter a few Puffins wander south to coastal and offshore waters of Massachusetts, and very rarely New Jersey. Early state records came from 1876, 1926, 1933 (three), 1951, and 1954. Today a few are recorded almost annually from winter pelagic trips (December–January).

PIGEONS AND DOVES / *Columbidae*

ROCK DOVE or DOMESTIC PIGEON *Columba livia* Gmelin

This European species was introduced to eastern North America long ago, and it is now widespread. It is particularly common about urban areas where it is often a nuisance. (Control measures are difficult and seldom complete.) There is a considerable mix of feral individuals with escapes from lofts. The species has a remarkably long breeding "season," from January into September or even December. Individuals can be triple brooded within an annual cycle, with two eggs in each clutch. (The flimsy stick nests are often constructed on a windowsill or ledge, and they afford little protection.)

High counts are often made at cornfields in the late fall (feeding flocks can contain several hundred).

BAND-TAILED PIGEON *Columba fasciata* Say

This western species has recently had a small number of extralimital records in the eastern United States. There is a single New Jersey report, and it can be listed only as hypothetical. The sighting was of a bird in flight on the Kittatinny Ridge* by two observers on September 17, 1980. As a cautionary note it should be recognized that the domestic pigeon has many plumage variations which include band-tailed patterns.

WHITE-WINGED DOVE *Zenaida asiatica* (Linnaeus)

This western species is accidental in northeastern states (e.g. there are several records from New York). New Jersey sightings were made at West Cape May on May 24, 1981, and in September 1983, but full verification of these reports was not possible. It should probably also be listed as a possible escape, since it is commonly kept in captivity. Hence, until further records it is only of hypothetical status.

MOURNING DOVE *Zenaida macroura* (Linnaeus)

The Mourning Dove is a permanent resident throughout the state. Formerly it was much less common north of the coastal plain, especially during winter. It has a prolonged breeding season (February–October), and may often have two clutches per year. Apparently some individuals are migratory, while others are sedentary through the year. The migratory periods are in March and September–October. During migration and through winter months large flocks (up to 1,000 birds) may be seen. The Mourning Dove is a popular game bird in many states, and hunters have suggested that it is common enough to list as such in New Jersey.

PASSENGER PIGEON *Ectopistes migratorius* (Linnaeus)

This large pigeon is now extinct. It was formerly abundant in the Appalachians, and migrants passed through our region (especially in the fall). It was very social with dense breeding colonies and immense migratory swarms (millions of birds). Apparently it was an abundant migrant as nearby as Pennsylvania (Wood 1979). The last reports from New Jersey include 2 collected near Englewood in September 1878, and a small flock of 10 near Morristown on October 7, 1893. No individuals have been seen in New Jersey since the turn of the century. (The last captive individual died in Ohio in 1914.)

COMMON GROUND-DOVE *Columbina passerina* (Linnaeus)

This little dove nests as far north as North Carolina but is only an accidental visitor to New Jersey. As its name suggests, it is generally quite terrestrial. Our records are from Camden (1858), Cape May (1935), and Brigantine National Wildlife Refuge* (fall 1961 into 1962). Similarly, there are only a few records from Pennsylvania and New York State.

PARROTS / *Psittacidae*

MONK PARAKEET *Myiopsitta monachus* (Boddaert)

The Monk Parakeet was introduced to the United States from its native South America by the pet trade. It has been actively controlled in many states as a potential agricultural pest (e.g. to fruit trees). It appeared in New Jersey in 1968, from a starting population on Long Island, New York. By the mid-1970s it had become commonly reported near various urban areas, including the Oranges and Camden. The state's elimination program has been fairly successful, but some breeding birds were still nesting in 1981 (e.g. June at Sea Bright). With control programs in many states it seems that the Monk Parakeet may have only a tenuous future in spite of its initial success (recorded in at least 38 states).

CAROLINA PARAKEET
Conuropsis carolinensis (Linnaeus)

This native North American parrot is now extinct. It was a common bird of the southeastern United States, but severe shooting greatly reduced its populations by the late 1800s. It is listed only hypothetically for New Jersey; there were unsubstantiated records from several summers in the 1850s at East Orange. There are old records from Maryland, Pennsylvania, and New York.

CUCKOOS AND ANIS / *Cuculidae*

BLACK-BILLED CUCKOO
Coccyzus erythropthalmus (Wilson)

The Black-billed Cuckoo is a migrant from its North American breeding range to its wintering grounds in northwestern South America. In our area it is both a migrant and a summer resident throughout, but it breeds more commonly in northern New Jersey. It nests along forest borders or in open woodlots, between its May arrival and August departure. (Rarely, the Black-billed Cuckoo has been noted to lay its eggs in nests of other bird species, such as the Catbird and Yellow Warbler. It is the European Cuckoo that is a nest parasite.)

YELLOW-BILLED CUCKOO *Coccyzus americanus* (Linnaeus)

The Yellow-billed Cuckoo breeds throughout the United States and winters in South America (as far south as Argentina). In New Jersey it is a summer resident, most numerous from May through September. Abundances vary from year to year, but it is often fairly common. It nests in thickets and second growth, often amid grape vines.

A high count of 34 comes from New Brunswick on May 13, 1952.

SMOOTH-BILLED ANI *Crotophaga ani* Linnaeus

This odd relative of the cuckoos may come northward from Florida only as an accidental visitor. There is one record, an 1849 specimen from Petty's Island,* Camden County, but Poole (1964) discounts this as a probable escape. Our neighboring states apparently have no records.

BARN-OWLS / *Tytonidae*

COMMON BARN-OWL *Tyto alba* (Scopoli)

This cosmopolitan bird is an uncommon permanent resident in our state, and it is particularly scarce in northwestern New Jersey. Some individuals, probably from the north, are migratory. The species is adaptable to urban habitats, and has been introduced (with nesting boxes) to city parks for rodent control. In our area it nests from March to September, with the majority raising an April–May brood. The Barn-Owl adopts a wide variety of man-made structures for nesting (e.g. church steeples, barns, and silos), and the bird is most easily encountered by searching such structures.

High counts are made during fall migration. For example, the Cape May Lighthouse* sometimes has many circle at night (maximum of 50 in one evening, with some possible duplication). Over 100 were banded at Cape May in the fall of 1977. (Barn-Owls banded in New Jersey have been recovered in Alabama and Florida [Soucy 1980b].)

OWLS / *Strigidae*

EASTERN SCREECH-OWL *Otus asio* (Linnaeus)

The Eastern Screech-Owl is a common permanent resident throughout our region of the United States. Both red- and gray-phased individuals are common in New Jersey. The species favors stands of small to moderate-sized trees, where it will nest between mid-March and mid-June. In addition to natural cavities, appropriate bird boxes are quickly adopted by this species.

Recent night surveys have used tape-recorded owl calls and have shown responding Screech-Owls to be remarkably numerous in some areas. In fact a new national high count for the Screech-Owl was obtained on a 1971 Moorestown area Christmas count, with 71 birds. (A summer high estimate of 70 was made at the Great Swamp National Wildlife Refuge* in 1974.)

GREAT HORNED OWL *Bubo virginianus* (Gmelin)

The Great Horned Owl nests throughout most of the western hemisphere, from subarctic Canada in the north to Tierra del Fuego in the south. In general it is a permanent resident of large woodlands. Some appear tolerant of human activity in large parks and urban woodlots. The Great Horned Owl is an unusually early nester, and it is often the first New Jersey bird on eggs each year (February to March). It frequently uses old hawk nests (e.g. of the Red-tailed Hawk) for its own nesting. As with most of our owls, the female does most of the incubation, and the male helps provide food. (Birds of the pale northern race, *wapacuthu,* have been very rarely recorded in nearby states, but there is no record from New Jersey to date.)

SNOWY OWL *Nyctea scandiaca* (Linnaeus)

The Snowy Owl is a rare winter visitor from the high arctic. It nests well north of the tree line, and can even winter at the edge of the Arctic Ocean. The irregular invasions into northern states during "flight years" are most frequently reflected in coastal records, particularly on dunes and at salt marshes. However this species is also recorded inland (e.g. Hightstown), and at rodent-infested landfills in urban areas. Most records are from mid-November through mid-March.

Recent invasion years have included the winters of 1954–1955, 1960–1961, and 1980–1981. Records from the 1960 flight came from Fair Lawn and Libertyville and from the 1980–1981 invasion from Sandy Hook, Island Beach,* Stone Harbor, and Brigantine National Wildlife Refuge.* (In 1981–1982 there was a light flight, with several records along the coast and two inland.)

It seems that two Snowy Owls at a time make a maximum: 2 on Brigantine Island* in late December 1964, 2 at Sandy Hook in November 1980, and 2 at Island Beach* from November 1980 into 1981.

NORTHERN HAWK-OWL *Surnia ulula* (Linnaeus)

This diurnal owl of northern muskegs is an extremely rare winter visitor as far south as New Jersey. Reportedly 3 birds were shot within the state during the mid-nineteenth century, but details are lacking. There are only two modern records: 1 shot in the winter of 1904 in Essex County, and 1 at New Brunswick on December 26, 1926. (Pennsylvania has a few sightings including 2 in 1963, and Long Island, New York, has two old records.)

BARRED OWL *Strix varia* Barton

The Barred Owl has a wide distribution in the United States and Canada. In our area it is a permanent resident of wet or rich woodlands. The species had declined in recent decades until a modest increase in numbers began at the end of the 1970s. Breeding areas have included the northwest corner of the state, the Great Swamp,* northern Cape May County, and the Delaware Bay shore of Cumberland County. It is most vocal in March, just before its April–May breeding season. It will often nest in a tree cavity, or an abandoned hawk nest (e.g. of a Red-shouldered Hawk).

A high count of 15 was reported at the Great Swamp National Wildlife Refuge* in the summer of 1974.

GREAT GRAY OWL *Strix nebulosa* Forster

This majestic owl is a western and northern species, breeding only as far east as Ontario. In invasion years individuals occur as very rare winter visitors to northeastern states (especially New Hampshire forests). For New Jersey it is considered hypothetical. There are two old (1859 and 1887) winter reports of birds shot in Sussex County and Morris County, respectively, but these have been questioned as details are lacking.

The species is an extremely rare winter visitor in Pennsylvania and New York (recently on Long Island), and it should occur in northern New Jersey sometime.

LONG-EARED OWL *Asio otus* (Linnaeus)

The Long-eared Owl is an uncommon winter visitant in coniferous stands, usually with several birds per roost. Rarely, it is a permanent nesting resident of coniferous or deciduous forests at scattered localities (e.g. Salem, Morris, Middlesex, Essex, Sussex, and especially Hunterdon counties). It nests in April and often adopts an old crow's nest. Most records of winter roosts are from evergreen groves within the period of November to March. (Roosts in our area may contain up to 18 Long-eared Owls.)

SHORT-EARED OWL *Asio flammeus* (Pontoppidan)

The Short-eared Owl is a worldwide species, even distributed on many oceanic islands. In our area it is a winter visitant, particularly along coastal marshes (November–March). It has declined substantially in recent decades, and it is but a rare local breeder (primarily on salt marshes of southern New Jersey). During the 1970s no nests were reported in the state, although several pair and an immature were seen in the summer of 1979.

Winter roosts in open country formerly were as large as 150–

200 owls! Today 10–20 is considered a high count (e.g. recent winters at Skillman).

BOREAL OWL *Aegolius funereus* (Linnaeus)

This aptly named little owl is an inhabitant of the northern (boreal) forests, and it is only an accidental visitor as far south as New Jersey. Our state has only one substantiated record, a bird at the Raritan Arsenal* on November 1, 1962. The bird was captured and subsequently died, and was prepared as a specimen (Newark Museum). A single observer report from Kingsland,* Bergen County, on March 12, 1929, has been questioned. (There is one record from Pennsylvania and a dozen reports from New York State.)

NORTHERN SAW-WHET OWL *Aegolius acadicus* (Gmelin)

This tiny owl is an irregular winter visitor in conifers (November to mid-March). Rarely, it is reported in summer as a probable breeding bird in cedar swamps of the Pine Barrens* in southern New Jersey. The species has bred in the Shawangunk Mountains of nearby New York, and it might be expected to nest at higher elevations of northwestern New Jersey as already indicated by the map in Peterson's *A field guide to the birds* (1980).

A great flight year was the winter invasion of 1978–1979, with many reports throughout the state. The winter of 1965–1966 was also a strong flight year, and 62 were reported along the Delaware River south of Camden! Summer birds have been noted at Lebanon State Forest* for several years, and nearby at New Lisbon (1968) and Chatsworth (1967). In northern New Jersey 5 were summer residents at the Great Swamp* in 1973. In the fall migration of 1981 a total of 110 Saw-whet Owls were banded at Cape May Point.

NIGHTJARS / *Caprimulgidae*

COMMON NIGHTHAWK *Chordeiles minor* (Forster)

The Common Nighthawk breeds throughout most of North America and winters throughout most of South America. It is more likely to be seen during daylight hours, especially at dusk, than other species in this nocturnal family. As a widespread migrant in our state, the Common Nighthawk is especially numerous in the fall (late August–early September). As a summer resident it is most common in urban areas, where it nests on rooftops. Elsewhere it is a scarce breeder (e.g. in recently burned areas of the Pine Barrens*).

High counts of migrants include 300 along the Delaware River (August 25, 1979), 500 at Elizabeth (September 22, 1936), and 575 at West Orange (August 1969).

CHUCK-WILL'S-WIDOW *Caprimulgus carolinensis* Gmelin

The Chuck-will's-widow is a summer resident of woodlands in the southeastern United States, although its breeding range has been gradually expanding northward (the first nesting in New York State was in 1975). In New Jersey it is an uncommon summer resident, May through August, primarily in coastal and southern areas. New Jersey populations became established in the 1920s at Cape May and have gradually increased. The first nesting was at Cape May Point (May 21, 1922), and local records continued with 20 singing birds in mid-June 1954. At the same time the species was gradually expanding along the coast, including Oceanville (as early as 1973) and Island Beach* (starting in 1978; more than a dozen in 1980). Marmora had nesting in 1968 and 1969, and singing birds were recorded at Sandy Hook in May 1978. The nesting locations are almost exactly the same from year to year, with two eggs laid on the ground amid dead leaves or pine needles.

In spring migration some individuals "overshoot" and are recorded north of the breeding range (e.g. in Massachusetts), and during this period of May movement Chuck-will's-widows are often heard at various nonbreeding sites scattered throughout New Jersey.

WHIP-POOR-WILL *Caprimulgus vociferus* Wilson

The Whip-poor-will is a migrant and summer resident, particularly in the Pine Barrens.* It is locally abundant as a nesting species in dry woodlands where two eggs are laid on the ground amid dead leaves. Nesting locales include the high areas of the Kittatinny Ridge,* where it has, however, declined in recent years.

The Whip-poor-will is most numerous during migration, but it is most frequently reported when calling on nesting territories. Its summer residence extends from May through August.

SWIFTS / *Apodidae*

CHIMNEY SWIFT *Chaetura pelagica* (Linnaeus)

The Chimney Swift breeds throughout the eastern United States and winters in the Amazon basin of Peru and western Brazil. In New Jersey the swift is a common migrant and summer resident. Formerly it nested in hollow trees, but today it is modernized and almost exclusively limited to nesting in chimneys (the stick nest is attached to the brick surface by the bird's glutinous saliva). The species is most numerous from late April through September, but stragglers can be seen into early November with mild falls.

High counts are made in early fall when the birds mass in large roosts, at chimneys. Reports include 3,000 at Montclair on September 2, 1940, and 4,000 at Ridgewood on September 13, 1976.

HUMMINGBIRDS / *Trochilidae*

RUBY-THROATED HUMMINGBIRD

Archilochus colubris (Linnaeus)

Our smallest bird is a migrant and summer resident throughout most of the state. (In winter it retreats to tropical areas.) In recent years the bird has been quite uncommon as a New Jersey breeder. The tiny cup nest, built by the female, holds two eggs and is often located in the same tree from year to year. Most arrive here in early May and depart by mid-September. Unfortunately our northeastern region has had a decline in hummingbirds during recent decades, and New Jersey recorded considerably fewer birds in 1980 and 1981.

KINGFISHERS / *Alcedinidae*

BELTED KINGFISHER *Ceryle alcyon* (Linnaeus)

The Belted Kingfisher breeds throughout most of North America. In New Jersey it is an uncommon migrant and local breeder. Some remain through the winter in areas of open water, especially along the coast in southern New Jersey. In part its nesting population may be restricted by limited nesting habitats (e.g. sandbanks and road cuts in sandy soil). The bird excavates a deep burrow with five to eight eggs laid directly on the tunnel floor.

Migration is considerably longer in the fall (August to late October) than in the spring (April to early May). Formerly the species was more common, and high counts during fall migration could tally a dozen or more per day along the coast.

WOODPECKERS / *Picidae*

RED-HEADED WOODPECKER
Melanerpes erythrocephalus (Linnaeus)

This eastern species has declined throughout much of its range. In our state it is now a rare migrant (May and September–early October) and a rare winter visitor. It is also a very rare breeding bird with three principal nesting areas (in northwestern New Jersey, in the Pine Barrens,* and in Cape May County). Its decline is assumed to be related to nest site competition with Starlings, and high mortality on highways (because of its low flight habits). In the past most nesting was in the northern part of the state; today most breeding records come from the south.

High counts include 15 breeding pair in Sussex County in 1960 and 10 adults at Cape May Court House in August 1977. Seven pair were found in the northern Pine Barrens* in the summer of 1980.

RED-BELLIED WOODPECKER
Melanerpes carolinus (Linnaeus)

This southern species was formerly very rare in the state (there were only five records from 1903 to 1935). As a nesting resident it was first restricted to Salem, Cumberland, and Cape May counties. It became more widespread through the 1960s (e.g. nesting at Princeton in 1962). By 1976 it was breeding in the northern areas (Sussex County), and 17 were recorded on the Hunterdon Christmas count on December 16, 1979. (The Red-bellied Woodpecker has now expanded as a breeder in New York State to southeastern counties and the Finger Lakes district.)

YELLOW-BELLIED SAPSUCKER
Sphyrapicus varius (Linnaeus)

The Yellow-bellied Sapsucker is the most migratory of our woodpeckers, breeding north of New Jersey and generally wintering to the south. It is a migrant in April and October, and rarely a winter resident (with only a dozen or so yearly on all of the state's Christmas counts). The species is widely distributed as a breeding bird in nearby southern New York State (e.g. Sullivan and Ulster counties), and it might sometime nest at higher elevations in northwestern New Jersey. (Searches for nesting birds should be made especially at forest borders of lakes.)

High counts during migration include 15 at Englewood on April 15, 1927, and 15 at Higbee Beach,* Cape May in mid-October 1981.

DOWNY WOODPECKER *Picoides pubescens* (Linnaeus)

This little woodpecker is a resident throughout most of North America. It will nest in almost any area where trees are available—it is adaptable to parks and suburbs as well as woodlands. Throughout New Jersey it is a common breeding bird, with families raised in April and May. The Downy Woodpecker is one of the most frequent visitors to suet at feeding stations, even in the summer. The species appears to be quite sedentary within the state, with almost no seasonal movement recorded.

HAIRY WOODPECKER *Picoides villosus* (Linnaeus)

The Hairy Woodpecker is found throughout most of North and Central America. (Tropical birds are smaller and darker than ours.) In New Jersey this woodland species is basically a permanent resident throughout, but it does exhibit some limited migratory movements and winter wanderings. Courtship begins early each year, and culminates with a clutch of three to six eggs by late April.

The Hairy Woodpecker is clearly less common than its smaller relative, the Downy Woodpecker (the ratio on New Jersey Christmas counts is often close to 1:5).

RED-COCKADED WOODPECKER *Picoides borealis* (Vieillot)

The Red-cockaded Woodpecker is an endangered species of the southeastern United States. Its populations are small and local, and declining in many areas. New Jersey has only one record, and this accidental species might not again occur here. The single record is of an old (prior to 1866) specimen from Hoboken (the specimen is now in the collection of the American Museum of Natural History,

New York City). The Red-cockaded Woodpecker has occurred in Maryland (bred) and Pennsylvania (prior to 1877).

THREE-TOED or NORTHERN THREE-TOED WOODPECKER *Picoides tridactylus* (Linnaeus)

In the East this species is the rarer of the three-toed woodpeckers. It is a sedentary species of northern woods (boreal forests) and is only accidental in our region. We have reports from only one winter (1958–1959), in northern New Jersey. Remarkably, 3 birds (1 male and 2 females) were seen on the Hackensack Christmas count of December 27, 1958. The birds foraged in pines at the Oradell Reservoir* and were seen repeatedly, by various observers. (Pennsylvania has had three records, all in 1974. It is a very rare breeding bird in the Adirondacks of northern New York.)

BLACK-BACKED or BLACK-BACKED THREE-TOED WOODPECKER *Picoides arcticus* (Swainson)

This northern species is a resident of boreal forests and breeds no closer than northern New York State. In New Jersey it is a very rare fall or winter visitor, with records from October to December.

Observations include Englewood on November 29, 1928, South Mountain Reservation* on October 15, 1950, Culver's Gap* on October 20, 1956, and Oradell Reservoir* on December 29, 1974, and the winters of 1956–1957 and 1958–1959. More recently birds were at Cape May on October 5–6, 1963 (2 males seen by many observers), Island Beach* on September 28, 1965 (banded and photographed), near Salem from late October to early November 1969, at Oradell Reservoir* on December 29, 1974, and Princeton, April 1983.

NORTHERN or COMMON FLICKER *Colaptes auratus* (Linnaeus)

This woodpecker nests throughout most of North America, in a wide variety of forests, woodlots, and orchards. In New Jersey it is a common migrant (mid-March to April and September to mid-October) and a summer resident throughout. It is relatively uncommon in winter, particularly when snow cover precludes its favored ground foraging. The nesting cavity of flickers is usually dead or partly diseased trees, but they sometimes use telephone poles or buildings.

PILEATED WOODPECKER *Dryocopus pileatus* (Linnaeus)

Our largest woodpecker is a permanent resident in wooded and mountain areas in the northern half of the state. As in other states its distribution is often local. However, it is adapting to human intrusion, and the species has probably been at its highest population levels in recent decades. It is generally quite rare in southern areas (e.g. the Pine Barrens*). Nesting begins in late April when a nest hole is excavated, usually in a tall dead tree.

A high count of 15 was totaled on the combined Christmas counts of the Great Swamp* and Boonton (winter of 1974–1975).

FLYCATCHERS / *Tyrannidae*

OLIVE-SIDED FLYCATCHER *Contopus borealis* (Swainson)

The Olive-sided Flycatcher breeds across Canada, and south into the United States at higher elevations. It breeds as nearby as the Catskills of New York, and in counties of north-central Pennsylvania. In 1954 it nested even closer to our state, at Orange County, New York, and occasional summering birds in the northern parts of our state suggest that we may have a New Jersey nesting record in the near future.

For now this species remains an uncommon to rare migrant (late May and early September). It is almost invariably seen perched near the top of a tall dead tree, from which it makes repeated sallies.

A high count of 4 was reported at Englewood on August 30, 1925.

WESTERN WOOD-PEWEE *Contopus sordidulus* Sclater

This western species breeds west of the Great Plains, to the Pacific coast. It is an accidental visitor to eastern states (e.g. Massachusetts, and three records from Maryland). There is a single New Jersey record of a juvenile male collected at Morristown on August 30, 1887 (the specimen is at the American Museum of Natural History in New York City). Bull (1964) cautions that the appearance of the two wood-pewee species is very similar, although the voices are very different and helpful for identification.

EASTERN WOOD-PEWEE *Contopus virens* (Linnaeus)

The Eastern Wood-Pewee breeds throughout the eastern United States and southeastern Canada. The winter season is spent in northwestern South America, often in sympathy with a tropical *Contopus* species.

In New Jersey the Wood-Pewee is a common migrant (May and September) and woodland breeder. The lichen-covered nest is usually built in oak trees by the female. Egg dates for our area range from early June to mid-July.

YELLOW-BELLIED FLYCATCHER *Empidonax flaviventris* (Baird & Baird)

The Yellow-bellied Flycatcher nests in coniferous forests of Canada, south to New York's Adirondacks (and rarely, the Catskills). It winters in Central America.

In New Jersey this little flycatcher is a rare migrant (late May and late August to mid-September). It is more frequently encountered during the fall migration when, alas, lack of vocalizations and the yellowish wash of some other *Empidonax* species can add to identification problems.

Island Beach* reported high counts of 6 on September 10, 1972, and a seasonal tally of 51 banded there during the fall of 1963. Aside from such banding station reports, it is very rare indeed to have more than one or two per migration at a given locality.

ACADIAN FLYCATCHER *Empidonax virescens* (Vieillot)

The Acadian Flycatcher breeds throughout all the eastern states, north to the latitude of southern New York. It winters in tropical forests from southern Central America to the equator. In New Jersey it is a local summer resident (May through mid-September); before 1900 it was more widespread as a breeding bird. Present breeding localities include Bull's Island State Park* (Delaware River), and scattered sites in northwestern counties and some areas in southern New Jersey (e.g. Cape May and Cumberland counties). In southern areas it is primarily found in hardwood swamps, whereas northern New Jersey populations favor hemlock ravines for a breeding habitat. Our northern breeding range has had a modest expansion in recent years (1978–1980) to include Big Flat Brook,* Pequannock, and Stokes State Forest.*

ALDER FLYCATCHER *Empidonax alnorum* Brewster

The Alder Flycatcher breeds from Canada south into northeastern states. It has decreased in some areas of its breeding range, including our area, because of the expansion of its close relative, the Willow Flycatcher. The Alder Flycatcher winters in tropical America.

In New Jersey it is an uncommon migrant and local breeder in wet areas with shrubbery. The nest is usually low (it averages less than two feet above the ground), in a shrub or small tree. Breeding sites are widespread with scattered individuals, mostly in northern areas (e.g. Stokes State Forest*). Central New Jersey has relatively few nesting birds, and in some years there may be none (e.g. irregular nesting at Princeton).

WILLOW FLYCATCHER *Empidonax traillii* (Audubon)

In the East the Willow Flycatcher nests in all states north of South Carolina. It has been expanding its breeding range in many areas, but it is still an uncommon migrant and local breeder in New Jersey. In recent decades its breeding range has expanded to include our coastal plain. Nests are in brushy thickets, either in old fields or at the borders of swamps. In our area the compact nest is usually in shrubs such as viburnum or elderberry.

Five pair apiece were reported at the Great Swamp,* and Great Piece Meadows* in the summer of 1979.

LEAST FLYCATCHER

Empidonax minimus (Baird & Baird)

The Least Flycatcher breeds across Canada and in northern states, and it winters in Central America. In our area it is now an uncommon migrant (May and late August to mid-September). It nests primarily in northern New Jersey, in open woodlands and orchards. The female alone incubates on the tiny cup nest (mid-May to mid-June). Our littlest flycatcher was formerly common, but it has recently been low in number. In fact, the population depression of the late 1970s saw its loss as a nesting species in many areas. (All four eastern *Empidonax* bred at the Stokes State Forest* area in 1977.)

WESTERN FLYCATCHER *Empidonax difficilis* Baird

This western species may be an accidental visitor to eastern states. There have been only two published reports, both from Brigantine National Wildlife Refuge.* The first listed without comment (!) a "Western Flycatcher" sighted on November 13, 1966; this may have involved an error in listing a Western Kingbird, but with the lack of details it is impossible to tell. The second, and perhaps our only good report, is of a November 16, 1981, sighting at the refuge. Still, the large yellow-bellied *Empidonax* only suggested that the bird might have been a Western Flycatcher, by appearance and date (see *American Birds,* 1982, 36:161). Although there are apparently no other records for this species in the Northeast, other western *Empidonax* have strayed eastward (e.g. *E. oberholseri* was collected in Pennsylvania in December 1969). Flycatchers of this genus seen in the late fall or winter should be carefully studied for possible western accidentals.

EASTERN PHOEBE *Sayornis phoebe* (Latham)

The Phoebe nests throughout much of the eastern United States except the more southerly areas. It winters in the South and in Mexico, although a few individuals remain as far north as our area. In New Jersey it is a common migrant (late March–early April and early October) and a summer resident (most numerous in northern counties). Although it is rarely present in winter there are usually several Christmas count reports in the state each year (mostly along the coast).

The Phoebe usually nests on ledges under bridges or on buildings. The compact nest usually holds five eggs, one of which is often that of a Cowbird.

A high count of 6 was made on the Rutgers campus in New Brunswick on March 18, 1982.

SAY'S PHOEBE *Sayornis saya* (Bonaparte)

The Say's Phoebe breeds throughout the western United States and Canada. It occurs in the eastern states as an accidental visitor. New Jersey sightings include Ocean County (January 22–28, 1956), Tuckerton (September 30, 1948), Brigantine Refuge* (October 2, 1960), Island Beach* (September 27, 1975, and October 15, 1977), and Oakland, Bergen County (September 24, 1977). There were other reports in 1977, an unprecedented invasion year. Both Pennsylvania and New York State have had similar sets of fewer than 10 records.

ASH-THROATED FLYCATCHER *Myiarchus cinerascens* (Lawrence)

The Ash-throated Flycatcher nests throughout the southern half of the western United States, and it occurs in the northeastern states only as an accidental visitor. On several occasions small *Myiarchus* flycatchers have been identified as this species in New Jersey, but field identification is very difficult. There are, for example, specimens of quite small individuals of the Great Crested Flycatcher (*Murray* 1971). The most recent New Jersey report comes from Higbee Beach,* Cape May, on September 27, 1981. (Details of this photographed bird, believed to be an adult, were published in *Cassinia,* 1982, 59:85.)

This species has been collected or photographed in Maryland (twice), New York (several times), and Rhode Island (several times).

GREAT CRESTED FLYCATCHER *Myiarchus crinitus* (Linnaeus)

The Great Crested Flycatcher is a common summer resident of wooded areas in the eastern half of the United States. It winters in the tropics of Central America and northern South America. In New Jersey it is present May through August, and less commonly through October. The nest is usually in a natural tree cavity, although bird boxes or abandoned woodpecker homes are also used. (The nest characteristically includes a few pieces of shed snakeskin among its other materials.) In our area egg dates range from mid-May through June.

BROWN-CRESTED or WIED'S CRESTED FLYCATCHER *Myiarchus tyrannulus* (Müller)

This large flycatcher is essentially a tropical species which has only been recorded in the Northeast as an accidental. New Jersey has a single sight record from Week's Landing,* Cape May, on December 24, 1978. Because field identification of the *Myiarchus* flycatchers can be difficult the species is listed here as hypothetical for New Jersey.

GREAT or GREATER KISKADEE *Pitangus sulphuratus* (Linnaeus)

The Kiskadee ranges throughout the Neotropics, and it has occurred in the eastern United States only as an accidental. The single New Jersey record was made by many observers of one at Cape May Court House on December 26, 1960. Ernest Choate, a local expert, has prepared an unpublished report giving details of this well documented sighting.

WESTERN KINGBIRD *Tyrannus verticalis* Say

The Western Kingbird is primarily a breeding bird of the western states. Most individuals migrate to Central America for the winter, but small numbers go to the Southeast (e.g. Florida). In our area the Western Kingbird is a rare fall migrant, September into November or even December. It is most frequent along the coast, and usually there are at least several sightings in New Jersey each year.

At other times of the year it is very rare, but there are extra-seasonal records (e.g. a January bird at Assunpink,* an April 6, 1957, record at Port Elizabeth, and a May 19, 1977, sighting at Sandy Hook).

EASTERN KINGBIRD *Tyrannus tyrannus* (Linnaeus)

The Eastern Kingbird breeds throughout much of North America, and winters in South America, from Ecuador to Bolivia. In New Jersey it is a common migrant (peaks in May and August) and a summer resident. Rarely, it will arrive in March, or remain into October. This flycatcher nests in trees that are adjacent to open areas (e.g. forest edges and orchards). The bird will defend its clutch very aggressively against any intruder.

A high count of 1,000 migrants was made at Cranbury on August 28–29, 1979. (The species regularly forms loose flocks during migration.)

GRAY KINGBIRD *Tyrannus dominicensis* (Gmelin)

This large kingbird is a tropical bird of open areas, nesting northward only to southern Florida. In northeastern states it is thus an accidental visitor, and New Jersey has had two sightings. The first came from Cape May Point on May 30, 1923 (several observers), the second from two observers at Point Pleasant (Ocean County) on September 18, 1946. (New York State has five records, all from Long Island. The most recent was after a hurricane in 1960.)

SCISSOR-TAILED FLYCATCHER *Tyrannus forficatus* (Gmelin)

This spectacular flycatcher nests in the south-central states. Most individuals winter in Central America, but some are found along the Gulf coast and southern Florida. For the northeastern states it is a very rare visitor. New Jersey records include: Mercer County (April 15, 1872), Cold Springs (September 2, 1941), Piscataway (June 29, 1958), Carteret (September 8, 1958), Cape May Point (October 21, 1961), and Avalon (September 7, 1965). More recently it has been seen at East Millstone (May 16, 1970), near Whiting (May 28, 1972), and at Cherry Hill (May 9, 1974), Cape May (October 28, 1975), Sandy Hook (May 16, 1977, and May 14–18, 1982), and Sharptown (June 5, 1981). The Scissor-tailed Flycatcher usually frequents open country (e.g. fields and roadsides). In contrast with the distinctive adults, short-tailed immatures may not be so easily identified.

FORK-TAILED FLYCATCHER

Tyrannus savana Vieillot

This flycatcher is found throughout much of Central and South America. It occurs only as an accidental visitor to northeastern states, and New Jersey has had more than its share of records. Sightings include Trenton in the autumn of 1900, Bridgeton in December 1820, Camden in June 1832 (J. J. Audubon), Cape May from November 1–3, 1939, West Orange on May 29, 1967, Cape May Point on August 23, 1968, Brigantine Refuge* on September 4, 1972, and Cape May on October 11, 1978. (There are a few records from New York State and Pennsylvania.)

LARKS / *Alaudidae*

EURASIAN SKYLARK *Alauda arvensis* (Linnaeus)

This European species, a great songster, was once introduced in various areas of North America, but it is now extirpated except on Vancouver Island, Canada. Skylarks were released in the New Jersey–New York metropolitan area in the late nineteenth century. The two introduced populations in New Jersey both failed (1883 and subsequently). There is a recent sight record from Wildwood Crest on July 18, 1977 (details published in *Cassinia,* 1982, 59:52), which may have involved an escaped cage bird.

HORNED LARK *Eremophila alpestris* (Linnaeus)

The Horned Lark breeds in open areas throughout most of North America. In our state it is a permanent resident as well as a migrant and winter visitant. Horned Larks are most common here in winter, with flocks of several hundred appearing in snow-drifted fields. The breeding distribution at the turn of the century was uncertain in New Jersey. By 1933 nesting was confirmed at Mount Holly, and in 1955 breeding records came from all coastal counties as well as some inland ones. By 1970 there were nesting reports through much of the state. Still, it is usually uncommon and local as a breeding species. (The race that breeds here is the more white-faced "Prairie Horned Lark.") The shallow cup nest is found on lawns, sandy lots, golf courses, and the like.

A high count of 15 nesting pair was reported from New Brunswick in 1950.

SWALLOWS AND MARTINS *Hirundinidae*

PURPLE MARTIN *Progne subis* (Linnaeus)

This large swallow nests in most states and winters in South America. In New Jersey it is an uncommon migrant and local breeder (where nesting boxes are provided). It is present from April into early September and nests from mid-May into early July. Nest boxes can contain many apartments, as Martins will nest in remarkably dense colonies. The species had a population decline in this century, but current numbers appear stable. It is certainly benefiting from the recent increase in martin birdhouses amid our landscape.

An old high count comes from a late summer roost containing 15,000 martins, in Gloucester County.

TREE SWALLOW *Tachycineta bicolor* (Vieillot)

The Tree Swallow breeds across southern Canada and the northern states. It generally winters in southern latitudes, but some remain north along the Atlantic coast to New Jersey, or even Long Island, New York. In our area it is a common migrant (April and late August to October), often abundant, with spectacular fall concentrations along the coast. The Tree Swallow is a local nester, especially where groups of nesting boxes are provided. (Elsewhere it will use a wide variety of tree cavities, such as old woodpecker holes.) Small numbers remain into December, or even overwinter along our southern coast.

Migration swarms in late summer and early fall can contain many thousands of birds. A high count of an estimated one million passed through Cape May Point, September 30–October 1, 1974.

NORTHERN ROUGH-WINGED SWALLOW *Stelgidopteryx serripennis* (Audubon)

The Rough-winged Swallow breeds throughout most of the United States and winters in tropical areas north to the Gulf Coast states. In New Jersey it is a migrant and summer resident, present from April through August. It nests locally from mid-May to mid-June throughout the state. The bulky nest of plant litter is most frequently built under bridges, at large lakes, or at rivers. It is generally a scarce species away from its breeding localities.

BANK SWALLOW *Riparia riparia* (Linnaeus)

The Bank Swallow is a wide ranging species: it breeds throughout much of North America, and it winters over much of South America. In New Jersey it is a migrant and a scarce local breeder. It is present from mid-April to mid-September, with active nests from mid-May to late June. Bank Swallows typically nest in colonies at gravel-pit sandbanks, with each pair making a deep nesting burrow. Currently, New Jersey has very few breeding colonies, in part because of the limited nesting habitat.

A high count of 1,000 comes from the Great Swamp* on August 20, 1979. (These birds were in the early stages of migration with the late summer massing that is so common in swallows.)

CLIFF SWALLOW *Hirundo pyrrhonota* Vieillot

The Cliff Swallow breeds throughout much of North America, including Mexico. It winters in central and southern areas of South America. It is generally a rare migrant (May and August) in New Jersey, although concentrations of several hundred are sometimes recorded at lakes or large rivers when abnormally cold weather causes a migration stall. The nest is an elaborate mud-walled chamber, usually placed under bridges or barn eaves. As a nesting species this swallow showed a marked decline in the Northeast from 1900 through the 1960s. In New Jersey it was reported as a common breeder in many areas (before the turn of the century), but the decline left us with only 13 pair breeding in the state by 1979. Fortunately, the Bull's Island State Park* colony (with 10 nests in 1979) showed remarkable expansion to 134 nests in 1982, while a Lambertville bridge had 104 nests (Welton 1983). Natural population growth has been aided by a state program of providing artificial nest chambers under Delaware River bridges. A nesting report for southern New Jersey came from Cedar Run Lake* in 1964.

A migrational high count of 500 was made at Boonton Reservoir* on May 14, 1952.

BARN SWALLOW *Hirundo rustica* Linnaeus

The Barn Swallow is a common summer resident throughout most of North America. It migrates far to the south to winter in lower Central America and South America. Migration peaks occur in May and August, and most breeding occurs between mid-May and early July. The mud nest is plastered to barn eaves, garage walls, beams under bridges, and the like.

High counts are made either during spring migration (e.g. 3,000 at Boonton Reservoir* on May 14, 1952) or in late summer with premigration assemblages (e.g. 3,000 at the Great Swamp* on August 20, 1979). The late summer congregations often contain several swallow species.

CROWS, JAYS, AND RAVENS *Corvidae*

GRAY JAY *Perisoreus canadensis* (Linnaeus)

This sedentary species of northern woods (boreal forests) nests as nearby as the Adirondack Mountains in New York, but it is only an accidental winter visitor in our area. There are unverified New Jersey reports from only a single winter, 1960. The sightings came from High Point* and nearby along the Delaware River, in early January, early April, and March 27, 1960. At least some of the observers were previously familiar with the species. (Pennsylvania has also had only a few sightings.)

BLUE JAY *Cyanocitta cristata* (Linnaeus)

The Blue Jay is a permanent resident and common migrant in the eastern United States. In New Jersey spring migration peaks in late April to early May, often with day after day of overhead flights. Fall migration is more prolonged (late September into October). The Blue Jay's nesting season in our area extends from mid-April through June. The Blue Jay is adaptable to woodlands, suburban areas, and towns (it will often nest in ornamental evergreen trees).

High counts of migrants include 2,000 over Park Ridge on May 9, 1960, and 2,868 over Woodbridge on May 6, 1952.

BLACK-BILLED MAGPIE *Pica pica* (Linnaeus)

The Magpie is probably a very rare visitor from western states, but most records of this species in eastern states are considered to involve escapes. (The species has frequently been kept in captivity, and pet escapes have often occurred in the past.) Some records of apparently "wild" birds include: Atlantic city (November 16, 1933), Bergen County (February 12, 1936), Sandy Hook (June 4, 1952, and May 17, 1953), West Milford (early September 1959), Englewood (March 19, 1960), and Wildwood (May 19, 1965).

New York State has had numerous records, and Pennsylvania had escapes establish a wild population in several western counties until 1962 (Wood 1979).

AMERICAN or COMMON CROW *Corvus brachyrhynchos* Brehm

This intelligent species is widely distributed in North America. In New Jersey it is a common resident, nesting throughout from mid-March to mid-May. The nest is a large basket of sticks, often high in an oak tree. The Crow is also common as a migrant (March–April and October–November), particularly in northwestern New Jersey. Large numbers are recorded at winter roosts (see high count below). In the past this species suffered severe persecution as an agricultural pest. Today it is becoming a backyard visitor even in housing developments.

A high count of 9,000 comes from Piscataway on February 6, 1977, at a roost.

FISH CROW *Corvus ossifragus* Wilson

The Fish Crow is essentially a species of tidewater areas and large rivers, north to Massachusetts. In New Jersey it is a permanent resident along the southern coast and Delaware Bay shore. Further north it is primarily a summer resident and migrant. It nests in solitary pairs or in small colonies. (In New Jersey most nest reports come from holly or cedar trees.) The Fish Crow's range has been expanding inland in many northeastern states, along main river systems (e.g. the Susquehanna in Pennsylvania and the Hudson River in New York State).

This species forms winter flocks like the American Crow but with much smaller numbers. A high count of 100 in a winter congregation was made at Shark River* on November 20, 1957.

COMMON RAVEN *Corvus corax* Linnaeus

The Raven ranges widely over North America, except in most eastern states. In New Jersey it is a very rare migrant and winter visitor, mostly in the northern counties. Migrant visitors are most frequently seen along the Kittatinny Mountains* (e.g. Sunrise Mountain* and Raccoon Ridge*), from late October into November.

The Raven was once a breeding bird in New Jersey, nesting locally along the southern coast (e.g. at Tuckerton and Cape May) until the end of the nineteenth century. Our last breeding birds vanished about 1905. (This species also has decreased dramatically as a breeding bird in remote areas of New York and Pennsylvania.)

Some recent reports of vagrants include birds at Franklin Lakes (October 22, 1960), Princeton (October 21, 1967, and February 21, 1972), New Lisbon (October 28, 1967), Bearfort Mountain* (April 24, 1982), and the Great Swamp* (May 6, 1982).

CHICKADEES AND TITMICE
Paridae

BLACK-CAPPED CHICKADEE

Parus atricapillus Linnaeus

The Black-capped Chickadee breeds throughout much of the northern half of North America. In New Jersey it is a common resident in regions to the north of Trenton and Perth Amboy. This chickadee will nest in either deciduous or coniferous forests. The cavity nests are often in a dead tree stump, and birds lay their clutch sometime between late April and mid-June.

There is some movement southward in winter, particularly with infrequent irruption years. (A Pine Barrens* breeding record from Lakehurst in June 1940 may have involved hybridization with the Carolina Chickadee.) There was a large invasion of southern New Jersey in 1961, when the Cape May Christmas count tallied 34 Black-capped Chickadees (in addition to the usual Carolina Chickadees).

High counts are often made on Christmas counts (e.g. 525 from the Hunterdon census on December 16, 1979). A summer high of 50 was recorded at South Mountain Reservation* on August 13, 1979.

CAROLINA CHICKADEE *Parus carolinensis* Audubon

The Carolina Chickadee occurs in the eastern United States, northward approximately to the southern edge of the Black-capped Chickadee's range. It is a common resident in southern New Jersey, breeding as far north as Mercer and Middlesex counties, where there is some range overlap with its northern relative. The Carolina Chickadee nests between mid-April and early June, in similar habitats and situations as the Black-capped Chickadee (even the eggs of the two species are indistinguishable). Where there is doubt of chickadee identification the call should be used to distinguish the species, but mixed songs are sometimes noted in central New Jersey. Field studies are being conducted on this overlap problem by workers in several states.

BOREAL CHICKADEE *Parus hudsonicus* Forster

This resident of northern coniferous forests is a very rare winter visitor in our area, appearing with infrequent irruptive years. It often appears at feeding stations or in foraging associations with Black-capped Chickadees.

An invasion year occurred in the winter of 1961–1962 when at least 35 were recorded in the state. Another invasion in the winter of 1969–1970 saw 8 reported on New Jersey Christmas counts.

Two high counts of 4 each are recorded: Princeton in the winter of 1916–1917, and the Clinton Road* area near Newfoundland on March 14, 1982.

TUFTED TITMOUSE *Parus bicolor* Linnaeus

The Tufted Titmouse is a resident throughout most of the eastern United States. Formerly it was much less common in northern New Jersey, but there have been rapid increases in recent decades (particularly 1950–1959), and winter survival is enhanced by the modern abundance of feeding stations. It nests in woodlands with large trees, with eggs as early as mid-April. (The female does all of the incubation, although the male will feed her.) Recent state totals from Christmas counts suggest that we may now have a relatively stable population size, and any further increases in the near future seem unlikely.

NUTHATCHES / *Sittidae*

RED-BREASTED NUTHATCH *Sitta canadensis* Linnaeus

The Red-breasted Nuthatch occurs throughout most of the United States as a winter visitor or breeding resident. In our area it is primarily a migrant, varying greatly in numbers from year to year. It is most numerous in the fall (September–October), and is generally uncommon in winter. A marked invasion occurred in the winter of 1980–1981 (e.g. there were "hundreds" at Washington Crossing State Park).

The Red-breasted Nuthatch is only a very rare breeding bird in New Jersey, with scattered nonsustained nesting localities in conifers (in extreme northern counties it may nest regularly). There are breeding records in central New Jersey (e.g. Princeton in June 1974, and as far south as Dividing Creek, Cumberland County, in 1977). The Pequannock Watershed* area has the best breeding population (a small number of pairs in 1979).

A daily high count was set at Higbee Beach,* Cape May, on September 21, 1980, when 223 were tallied on migration.

WHITE-BREASTED NUTHATCH *Sitta carolinensis* Latham

The White-breasted Nuthatch is a resident throughout most of the United States. In our state it is a northern resident, nesting in woodlands. It becomes less common further south and is only a scarce migrant and winter visitant at Cape May. There are breeding records from Camden and Burlington counties, and parts of the Pine Barrens.* It nests in April–May using any of a wide variety of tree cavities. Winter populations vary but they are often considerably larger than summer ones, because of an influx of northern individuals. A Great Swamp*–Watchung Ridges Christmas count tallied a high of 205 on December 20, 1981.

BROWN-HEADED NUTHATCH *Sitta pusilla* Latham

This species of the pinelands in southeastern states may be an accidental visitor to New Jersey. The best New Jersey record came from the winter of 1876 with one at a suet feeder (!) in Haddonfield. There is an unspecific mid-nineteenth-century inclusion in a list for Cape May County birds, and a doubtful report from Hopatcong from the fall of 1968.

CREEPERS / *Certhiidae*

BROWN CREEPER *Certhia americana* Bonaparte

At one season or another the Brown Creeper occurs throughout the United States and southern Canada. In our area it is a migrant (April and October) and an uncommon winter resident. It is a scarce local nester in scattered localities in northern New Jersey and in hardwood swamps of the Pine Barrens.* Breeding locales in southern New Jersey in 1980 included parts of Ocean, Burlington, Atlantic, and Cumberland counties. The first New Jersey nesting was reported at Sussex County in 1906. There have been infrequent breeding records from central New Jersey (e.g. at Princeton in 1954 and 1980). Compact nests are made behind loose bark, between early May and early June. Small breeding groups have been reported from the Great Swamp* (7 in 1979) and Watchung Reservation* (6 pair in June 1981).

The species can be a common migrant along the coast; 2,600 were banded at Island Beach* in the fall of 1968.

WRENS / *Troglodytidae*

CAROLINA WREN *Thryothorus ludovicianus* (Latham)

The Carolina Wren is a species of the eastern United States, northward to southern New England. In New Jersey it is a permanent resident throughout, except in extreme northern areas. The species is subject to cyclic population changes, apparently related to high mortality in harsh winters. Still this species is by far the most common wren on our state's Christmas counts. Generally this southern bird is uncommon in our region and well restricted to thickets. The Carolina Wren's breeding season is long (April to August) and it is often double brooded at our latitude. The cavity nest is constructed in a wide variety of natural or man-made niches.

BEWICK'S WREN *Thryomanes bewickii* (Audubon)

The Bewick's Wren is widespread in western and southern states. It has been only a very rare visitor in New Jersey, but the species has been expanding its breeding range in neighboring states (e.g. southwestern counties of Pennsylvania and Ulster County, New York).

Records include Haddonfield (1890), Ho-Ho-Kus (May 24, 1935), and Sunrise Mountain,* Sussex County (April 20, 1958). More recently there have been birds at Cape May Point (October 13, 1962, and June–July 1970) and Island Beach* (October 11, 1969, and May 4, 1977, both banded).

HOUSE WREN *Troglodytes aedon* (Vieillot)

The House Wren breeds throughout most of the United States (the species also has a South American population). In our area it is a common summer resident (May through September), nesting throughout. It will have two broods between early May and mid-August. Nests are made in a wide variety of cavities, and it is especially fond of birdhouses.

Rarely, individuals remain into late fall, and there are a few reports of House Wrens each winter in New Jersey.

WINTER WREN *Troglodytes troglodytes* (Linnaeus)

This little wren nests in northern forests throughout North America and Eurasia. In New Jersey it is an uncommon migrant (April and October) and a scarce winter resident. It is a rare local nester in a few localities of northern New Jersey. (The nest is very well hidden amid tree roots or fallen stumps.) The first definite New Jersey nesting came from Lake Girard,* Sussex County, in June 1962. The Pequannock Watershed* area had a modest nesting population in 1979.

A remarkable high count of 22 came from Stokes State Forest* on April 22, 1944.

SEDGE WREN or SHORT-BILLED MARSH WREN

Cistothorus platensis (Latham)

The Sedge Wren nests in the temperate zones of eastern North America and South America. In our area it is a very rare migrant and very rare breeder (few if any nesting localities remain in New Jersey today). In the past it nested in both the northern and southern parts of the state. Our colonies occurred in freshwater meadows, inland marshes and bogs, and short-grass salt marshes. Possibly none was nesting in New Jersey as of 1980.

A nesting colony of 8 pair occurred at Troy Meadows* in 1947. Winter records have recently been obtained at Brigantine National Wildlife Refuge* (up to 6 birds).

MARSH or LONG-BILLED MARSH WREN

Cistothorus palustris (Wilson)

The Marsh Wren breeds across central North America. In New Jersey it is a common summer resident, nesting in both fresh and saltwater marshes throughout. It nests colonially, usually in stands of cattails, *Phragmites* reed, or small shrubs. It is present from late April through October, and uncommonly into late fall. Rarely, it remains into winter (mostly along the coast).

A high nesting count of 160 pair was obtained at the Troy Meadows* marsh in the summer of 1947. It has been found to be the most numerous breeding bird in cattail marshes in various areas, such as the Great Swamp* and the Trenton Marshes.* Censuses are easily conducted of singing birds, but they are often quite difficult to see.

KINGLETS, GNATCATCHERS, AND THRUSHES
Muscicapidae

GOLDEN-CROWNED KINGLET

Regulus satrapa Lichtenstein

As a breeding bird the Golden-crowned Kinglet is mostly associated with coniferous forests in southern Canada. In New Jersey it is usually a common migrant, particularly during fall migration (October). In winter it is a resident in evergreen groves. Numbers fluctuate from year to year, as reflected by large variations in the state Christmas count totals.

Rarely, the Golden-crowned Kinglet is a local breeder at scattered localities in extreme northern New Jersey (e.g. Pequannock Watershed* since 1971), again in conifer stands. The first New Jersey nesting was relatively recent: in 1963 in the Greenbrook Sanctuary* in Bergen County. The nest is typically a ball of hanging moss and lichens, and is constructed entirely by the female.

A seasonal high total of 2,419 was banded at Island Beach* during the fall migration of 1967.

RUBY-CROWNED KINGLET

Regulus calendula (Linnaeus)

The Ruby-crowned Kinglet is a common breeding bird of North America's northern coniferous forests. It breeds as close as the Adirondack Mountains of New York and "probably" in northern counties of Pennsylvania (Wood 1979). In New Jersey it is a migrant (April and October) and winter resident (mostly southern New Jersey and the coast). It is usually less common than the Golden-crowned Kinglet.

A high count of 100 was made at Sandy Hook during a migration peak on October 10, 1953.

BLUE-GRAY GNATCATCHER *Polioptila caerulea* (Linnaeus)

The Blue-gray Gnatcatcher is predominantly a southern species, although it has been expanding its breeding range northward through New England. It winters in southern states and Central America. In New Jersey it is a summer resident (mid-April through August), nesting in open woodlands throughout. Formerly this species was rather restricted to breeding in southern New Jersey. With two major expansions (1947 and 1954 flight years) and subsequent increases it has become fairly common in many areas of northern New Jersey as well. (It nested at High Point State Park* in 1960.) The nest is built in late April or May; it is a tidy little cup with lichen pieces on the outside.

A high count of 12 came from Sandy Hook on September 23, 1978.

NORTHERN WHEATEAR *Oenanthe oenanthe* (Linnaeus)

The Wheatear is a very rare visitor from the high arctic of Canada (it migrates to winter in the Old World). Almost all New Jersey records are from the fall. Sightings come from the coast at Cape May (October 1951, November 7, 1951, September 23, 1978, and September 27–28, 1981), and Brigantine Refuge* (two weeks in September 1970, and October 1, 1974). A curious Christmas count report came from Walnut Valley on December 20, 1980 (odd location and date). Our first New Jersey spring record came from Thompsons Beach in Cumberland County on June 1, 1981. Fall dates (September and October) dominate the reports from our neighboring states as well.

EASTERN BLUEBIRD *Sialia sialis* (Linnaeus)

The Eastern Bluebird nests throughout the eastern United States (westward into Arizona). In our area it is an uncommon migrant (March and October to mid-November), and a scarce winter resident. New Jersey also has local summer residents, especially where colonies adopt nesting boxes set out along "Bluebird trails." Formerly the species was more widespread and relatively common (e.g. about orchards). Some examples of notable breeding-bird success come from the Great Swamp* (with 100 in a nesting colony) and the Colliers Mills Wildlife Management Area* (47 young fledged in 1979).

Migration counts include 75 at the Wharton Tract* in the Pine Barrens* (October 19, 1959), 150 at Raccoon Ridge* (October 1977), and 440 at Cape May (October 25, 1935). A recent increase in the species is reflected by a remarkable total of 750–1,000 at Cape May on October 28, 1981.

High winter counts often come from north-central New Jersey (e.g. the Somerville Christmas count had 67 in 1959).

MOUNTAIN BLUEBIRD *Sialia currucoides* (Bechstein)

This species of western North America occurs only accidentally in our region. The single New Jersey record is that of a male at Brigantine National Wildlife Refuge* on November 21, 1982 (photographed). It foraged at the grassy edge of a pine woods. (The first state records for New York and Pennsylvania were made in 1974.)

TOWNSEND'S SOLITAIRE *Myadestes townsendi* (Audubon)

The Solitaire is an uncommon breeding bird of western coniferous forests. There is but one well substantiated record for New Jersey, and it is only an accidental visitor in any of the northeastern states. Our record comes from Sunrise Mountain* at Stokes State Forest* in the late fall of 1980. The bird arrived on November 17 and remained until December 14; it was seen on many dates in this interval, by a total of more than 50 observers. On most days it fed briefly on mountain-ash berries at the summit. (Another possible Townsend's Solitaire was reported by a single observer at Mt. Tabor on October 19, 1982.)

VEERY *Catharus fuscescens* (Stephens)

The Veery breeds in forests across the northern United States and southern Canada. Most Veeries winter in northwestern South America. In New Jersey it is a common migrant (May and September) throughout, and a summer resident in wet woodlands of the northern half of the state. It is probably a scarce breeder in the central and southern parts of the state, as summering birds are regularly reported (e.g. at Assunpink* and Turkey Swamp,* Monmouth County).

A high count of 92 was reported from Cape May on September 7, 1935.

GRAY-CHEEKED THRUSH

Catharus minimus (Lafresnaye)

The Gray-cheeked Thrush breeds in boreal forests across Alaska and northern Canada. It winters from lower Central America through northwestern South America. In New Jersey this thrush is a distinctly uncommon migrant, with a brief spring passage in mid-May. Fall migration extends from mid-September to mid-October. A migrational high count was made at Island Beach State Park* on September 27, 1963, when 53 were banded.

SWAINSON'S or OLIVE-BACKED THRUSH

Catharus ustulatus (Nuttall)

The Swainson's Thrush breeds in spruce forests across northern North America, and it winters from Mexico south to northern Argentina. It is a quite common migrant in our area, within forest tracts. Its spring migration passage (mid-May) is rather rapid. Fall flights are heavier and extend from mid-September to early October. During fall migration large numbers may be seen along the coast (e.g. 700 at Cape May Point on October 1, 1974).

HERMIT THRUSH

Catharus guttatus (Pallas)

The Hermit Thrush breeds in northern woodlands across North America. Most individuals winter in southern states or Central America, but some are found as far north as coastal Massachusetts. In New Jersey it is a widespread migrant (April and October) and an uncommon winter resident (most common along the coast). It is a scarce breeding species at scattered localities in northern New Jersey, although it is sometimes locally common (e.g. Stokes State Forest* in 1982). In 1981 it was also discovered as a probable breeder in southern New Jersey at three Pine Barrens* localities in Wharton State Forest,* Burlington County. The Hermit Thrush has a single clutch between mid-May and mid-June. (The ground nests will contain three to four eggs.)

WOOD THRUSH *Hylocichla mustelina* (Gmelin)

The Wood Thrush breeds in the eastern United States, and it winters in Central America. In New Jersey it is a common summer resident, particularly in rich woodlands of central and northern areas. In recent decades the species has been increasing its range northward in New England, and generally increasing in numbers as a breeding bird in southern New Jersey (within cedar swamps). It is present May to mid-October, and nests from May to early August (often with two broods). The nest is usually constructed on a low limb, and frequently it contains a few conspicuous pieces of paper or rag. Very rarely, single birds are reported in winter months.

A high count of 77 was tallied at Blairstown in the summer of 1966.

AMERICAN ROBIN *Turdus migratorius* Linnaeus

The Robin breeds throughout almost all of North America. In winter most go to areas of the southern United States and Mexico, but some are found as far north as Canada. In New Jersey the Robin is an abundant migrant (mid-March to May and October to mid-November), and widespread as a common nesting bird. Two broods can be raised between mid-April and late August. The familiar nest is started with a substantial mud cup placed in a shrub, tree, or building niche. Generally Robins are uncommon in winter, but in some mild years winter flocks and roosts can contain hundreds of birds each.

A high count of 9,815 was reported at Cape May on October 24, 1935.

VARIED THRUSH *Ixoreus naevius* (Gmelin)

The Varied Thrush is a breeding bird of the Pacific Northwest. It occurs as a very rare visitor to the eastern United States, usually in winter and often at feeding stations. For New Jersey there is a specimen of a male from March 1848 (locality?) and a December 1851 specimen from Hoboken. A male remained at Clementon from November 26, 1936, to March 20, 1937, a remarkable stay. More recently a male occurred at Moorestown from November 1965 to February 1966, and a possible mated pair occurred in northwestern

Sussex County from December 1975 to January 1976. In 1980 there were multiple records from northern New Jersey: Allamuchy Mountain State Park* on November 15–16, Sunrise Mountain* on December 2, and near Medford from December 17 to 21. This rare thrush has been recorded more frequently in our larger neighboring states: Pennsylvania has had at least 10 records (Wood 1979), and New York has had almost 30 (Bull 1974).

MIMICS / *Mimidae*

GRAY CATBIRD *Dumetella carolinensis* (Linnaeus)

The Catbird breeds throughout much of the United States, but in winter it generally retreats to warm climates of southeastern states and the tropics. In New Jersey it is a very common migrant and common summer resident, nesting in thickets and woodlands throughout. Most breed in May and June, with deep blue eggs set in a bulky but well-constructed nest. The Catbird is common from May into early October, uncommon through the remainder of the fall, and rare in mid-winter (usually at feeding stations).

A high count of 400 migrants was obtained at Cape May Point on October 1, 1974.

NORTHERN MOCKINGBIRD *Mimus polyglottos* (Linnaeus)

The Mockingbird is most abundant in southern states, but it has expanded its range into New England and even southern Canada in recent decades. It is a resident throughout New Jersey, most common in southern and central areas. It is common at housing developments, and in old fields (especially where there are multiflora rosebushes). A pair will often raise two broods between early April and August.

There were some nesting records in New Jersey in the 1880s (e.g. at Tenafly), but the species was extirpated as a breeding bird by the end of the century (nestlings were often taken as cage birds).

By 1955 it had become a local permanent resident in southern New Jersey, breeding as far north as Mercer and Middlesex counties. Nesting in extreme northern New Jersey came in 1958 at Sussex County (there were several here the next year).

A high count of 180 was obtained on the Great Swamp* Christmas count, 1977.

SAGE THRASHER *Oreoscoptes montanus* (Townsend)

The Sage Thrasher lives in arid areas of western states. It has been recorded in at least six eastern states, including New Jersey, as an accidental visitor. Our single sight record came from several observers at Barnegat Inlet* on November 27, 1949. New York State has had at least four records (most recently in January 1973).

BROWN THRASHER *Toxostoma rufum* (Linnaeus)

The Brown thrasher breeds throughout most of the eastern United States, and it winters, uncommonly, north to New Jersey. In our area it is a common migrant and summer resident (mid-April to September), nesting in thickets and shrubby fields throughout. The twiggy nest is often in shrubs or bushes (e.g. multiflora rose). The thrasher is uncommon or rare at other seasons (mid-winter reports are frequently at feeding stations).

High counts include 40 reported from Cape May Point on October 1, 1974, and an estimated 200 nesting pair at Sandy Hook in June 1976.

PIPITS / *Motacillidae*

WATER PIPIT *Anthus spinoletta* (Linnaeus)

The Water Pipit breeds at higher elevations and more northerly latitudes throughout much of North America. It winters in Central America and southern states, north to Long Island, New York. This migrant visits open fields throughout New Jersey in March to mid-

April and again from October to early November. It is a scarce winter resident, usually along our coast.

High counts include 125 at Assunpink* in April 1976.

SPRAGUE'S PIPIT *Anthus spragueii* (Audubon)

The Sprague's Pipit is a bird of central North America, breeding in prairie areas. It is only an accidental visitor to northeastern states. New Jersey has a single probable sighting published, for a bird at Brigantine National Wildlife Refuge* on May 10, 1959, by 21 observers of the Delaware Valley Ornithological Club. Wood (1979) reports on two sightings from western Pennsylvania, in 1957 and 1974.

WAXWINGS / *Bombycillidae*

BOHEMIAN WAXWING *Bombycilla garrulus* (Linnaeus)

The Bohemian Waxwing is a bird of northern forests, breeding in northwestern North America. Normally it winters in western states, but a few come east as very rare winter visitants. Bohemian Waxwings are found within winter flocks of Cedar Waxwings, so comparison is often easy. Most New Jersey records of this larger species are from "flight years" when the species expands southward and eastward of its normal winter range. In many years there are no state sightings.

Records include one at Princeton (February 1962), one at Flemington (April 10–22, 1962), and two at Barnegat Light (November–December 1968).

CEDAR WAXWING *Bombycilla cedrorum* Vieillot

The Cedar Waxwing breeds across central North America, and it winters throughout much of the United States and Mexico. In New Jersey it is a common migrant (May–early June and mid-August through October). The Waxwing is erratic in occurrence at all other seasons as it is a wandering nomadic. It is a relatively late nester, with a pair commonly producing two broods between early June and late August. Nests are placed in medium-sized trees at the forest edge, in open woods, in yards, and in orchards.

A high count comes from Ramsey where a winter flock contained 1,000 birds in February 1947.

SHRIKES / *Laniidae*

NORTHERN SHRIKE *Lanius excubitor* Linnaeus

This large shrike is found in both North America and the Old World. Its New World breeding range extends from Alaska across northern Canada. In our area it is a rare winter visitor, mostly in "flight years" with invasions from the north. On the East Coast the normal southern limit of such invasions is about the Delmarva Peninsula. Winters with pronounced flights included 1926–1927, 1930–1931, 1949–1950, and 1976–1977. Fortunately for birders some shrikes become dependable winter residents that can be seen over a period of weeks or even months (e.g. at Rosedale Park,* Princeton).

LOGGERHEAD SHRIKE *Lanius ludovicianus* (Linnaeus)

The Loggerhead Shrike breeds throughout most of the United States, but it becomes common only in the South. There are almost no breeding areas in northeastern states. In New Jersey it is a very rare spring migrant (mid-March to early April) and a very uncommon fall migrant (September–October). It is scarce in winter and is usually found in southern New Jersey (e.g. Delaware Bay shore). Formerly the species was more numerous. (It has bred in nearby parts of New York State, and western Pennsylvania, but we have no records of nesting in New Jersey.)

A high count of 6 comes from Cape May on September 16, 1934.

STARLINGS / *Sturnidae*

EUROPEAN STARLING *Sturnus vulgaris* (Linnaeus)

The Starling is native to the Old World, but it has been widely introduced in other parts of the world. In New Jersey it is especially abundant in urban areas and other developed habitats. It has probably been an important competitor (for food and nesting cavities) with native species. Usually two broods are raised between early April and mid-June. The nest can be placed in any of a wide variety of natural or man-made cavities, from woodpecker holes to bird boxes.

The Starling was introduced at Central Park in New York City in 1890, and now its range covers the United States. It entered New Jersey at Plainfield in 1900, and extended to Princeton (1905), Morristown (1907), and Cape May (nesting in 1909).

Counts of hundreds of thousands were made at the Raritan River* winter roosts in the 1930s. The roosts are still extant but reduced in size.

VIREOS / *Vireonidae*

WHITE-EYED VIREO *Vireo griseus* (Boddaert)

The White-eyed Vireo is a southerly species, breeding northward through New Jersey and along the coast to southern New England (it has been expanding northward in recent decades). In our state it is a summer resident (May–September), nesting in areas of thickets, in both swampy woodlands and old fields. Typically the nest is located in a shrub or low tree, and it is often victimized by Cowbird parasitism. The breeding season extends from early May to mid-July, and both sexes participate in incubation.

This species is decidedly more common below central New Jersey, and it was reportedly more common before the turn of the century.

BELL'S VIREO *Vireo bellii* Audubon

The Bell's Vireo is typically a breeding bird of western and midwestern states, preferring streamside habitats. It breeds as far east as Illinois and western Tennessee, but it is only an accidental visitor to the Northeast. New Jersey reports include an unsubstantiated sight record from Gloucester County on August 10, 1940, a female netted and collected at Island Beach* on September 15, 1959 (specimen at the American Museum of Natural History in New York City), a regional editor's sighting at Sandy Hook on September 4, 1978, and one at Maplewood on May 13, 1980 (two observers). There are only a few records from New York State and Pennsylvania, in summer and fall (although there are some unaccepted sightings from spring).

SOLITARY VIREO *Vireo solitarius* (Wilson)

The Solitary Vireo nests in mixed conifer and deciduous forests of higher elevations or more northerly latitudes. (New Jersey is along the southern boundary of its breeding range in the East.) It winters in tropical America.

In our area it is essentially a migrant (rare in the southern coastal plain), usually from mid-April to early May and late September to mid-October. It is a rare breeder in our state, primarily at higher elevations in Warren, Sussex, and Passaic counties. In this region the Solitary Vireo usually builds its nest in a hemlock or pine in early June.

YELLOW-THROATED VIREO *Vireo flavifrons* Vieillot

The Yellow-throated Vireo breeds throughout almost all of the eastern United States, and it winters in the West Indies and Central America. In New Jersey it is a summer resident (May–September), nesting in forested areas of northern New Jersey, and rarely in swampy woodlands of southern New Jersey. (Breeding in the southern part of the state was first reported in the 1970s, primarily in Cumberland County.) Its breeding season extends from mid-May into July. The nest is usually built in an oak or maple, close to the tree's trunk. Like the Red-eyed Vireo, this species has shown a marked decline in much of the state during recent decades.

WARBLING VIREO *Vireo gilvus* (Vieillot)

The Warbling Vireo breeds throughout most of North America where there are deciduous or mixed woodlands. It winters in Central American oak forests. This vireo has a very local distribution in many areas, and away from its breeding localities it is often a scarce migrant (late April–May and September–October). Very rarely, it is recorded in much of the coastal plain (e.g. Cape May). It is a regular summer resident, however, in the northern half of New Jersey, and parts of Burlington and Camden counties. It breeds in various shade trees, often near water courses. The compact cup nest is usually high in a deciduous tree, and both sexes incubate. It has been suggested that the decline of this species in the twentieth century might be associated with the Dutch elm disease that eliminated one of this vireo's favored nesting trees.

PHILADELPHIA VIREO *Vireo philadelphicus* (Cassin)

The Philadelphia Vireo breeds across southern Canada, and it winters in lower Central America and adjacent South America. In our area it is a rare fall migrant (mid-September) and a very rare spring migrant (May). It generally has a short period of passage, but in favorable spots it occurs annually.

High counts were obtained twice at Island Beach* where 9 were banded on both September 11 and September 12, 1960.

RED-EYED VIREO *Vireo olivaceus* (Linnaeus)

The Red-eyed Vireo breeds in forests throughout much of North America, and it winters in South America. (There is also a South American breeding population.) The species has been one of the most abundant birds in the eastern deciduous forests, and in our area it is a common migrant and summer resident (May–September). The breeding populations in New Jersey and other northeastern states appear to have decreased considerably in recent decades, possibly because of widespread forest fragmentation (Galli, Leck & Forman 1976). Red-eyed Vireos have a long nesting season (early May into August), and sometimes they will have two broods. (The deep cup nest is well attached to a tree fork.)

A high count of 100 was obtained at the New Jersey Audubon Scherman Sanctuary* during mid-September 1973.

WOOD WARBLERS / *Parulinae*

BLUE-WINGED WARBLER *Vermivora pinus* (Linnaeus)

The Blue-winged Warbler has a disjunct breeding distribution in the northern half of the eastern United States. (The range has been gradually increasing in recent decades.) It winters in Central America. As a New Jersey migrant and summer resident (May–August), it breeds in second-growth fields and woodlands (most commonly in our northern counties). Generally the eggs are laid in the first half of June, and the female alone incubates. (See comments about hybridization of this species under Golden-winged Warbler.)

GOLDEN-WINGED WARBLER *Vermivora chrysoptera* (Linnaeus)

The Golden-winged Warbler's breeding range has included areas of the northeastern United States, south to the Smoky Mountains. It has had some recent regional decreases, however, and it has been lost as a local breeder in many areas. In New Jersey it is an uncommon migrant and scarce summer resident (May–August). Nesting localities are primarily in northwestern counties, in second-growth fields. The bulky grass nest is built near to or on the ground, with four to five eggs laid in late May or early June.

Hybrids of this species with the Blue-winged Warbler are regularly reported in the northern half of our state. The "Brewster's" form, whitish, is decidedly more frequent than the "Lawrence's" form. (Both hybrid types have been reported breeding with parental species, albeit rarely.)

TENNESSEE WARBLER *Vermivora peregrina* (Wilson)

The Tennessee Warbler breeds across Canada and winters from Central America into South America. (The nearest nesting localities are in the Adirondacks of New York.) In New Jersey this migrant (May and September to early October) varies in abundance from year to year, but usually it is uncommon in spring and moderately common in the fall. Because of its treetop habits and relatively dull plumage it is more likely to go unnoticed than its brighter relatives.

ORANGE-CROWNED WARBLER *Vermivora celata* (Say)

The Orange-crowned Warbler breeds in the western United States and across Canada. Most winter in Central America and the southern states, but some stay northward along the Atlantic coast to Virginia or even New York. In New Jersey it is a very rare migrant, mostly reported in the fall (October), although spring sightings are recorded. Preferred habitats include old fields in the interior and bay-shore shrubbery along the coast. It is also very rare in winter months (individuals sometimes frequent feeding stations). Often there are no New Jersey reports on Christmas counts.

A high count of 11 was obtained at Cape May on December 26, 1956.

NASHVILLE WARBLER *Vermivora ruficapilla* (Wilson)

The Nashville Warbler has a disjunct breeding range in the eastern and western parts of northern states and southern Canada. It winters in Central America, north to the Gulf coast of Texas. In our area it is a migrant (early May and September–early October) and rare local breeder. It is reported nesting only from scattered localities in northern New Jersey (late May into June). The ground nest is hidden in mosses of bogs or forests.

VIRGINIA'S WARBLER *Vermivora virginiae* (Baird)

The Virginia's Warbler breeds in western states, and it is only an accidental visitor to eastern North America. There are two New Jersey records, both of birds banded at Island Beach State Park* (October 6, 1962, and September 24, 1966). (There are almost no other records from eastern states.)

NORTHERN PARULA or PARULA WARBLER *Parula americana* (Linnaeus)

The Parula breeds throughout eastern North America, and it winters in tropical America including the West Indies. In our area it is a common migrant (May and September to early October). Formerly it was a local breeder in southern New Jersey at coastal cedar swamps, and more rarely at bogs in northern areas of the state. It was extirpated as a breeder in our state in the 1930s, apparently coincidentally with the decline of *Usnea* lichen that is used in its nest construction.

There was a nesting record from Batsto in the Pine Barrens* (1958), and summer singing birds have been reported in Cumberland and Ocean counties (1977, 1978, and 1980). Currently the nesting status in most areas is uncertain, but in 1983 it was apparently breeding again along the Delaware River.

YELLOW WARBLER *Dendroica petechia* (Linnaeus)

The Yellow Warbler breeds throughout almost all of North America, and it winters widely in the West Indies, Central America, and northern South America. (It has considerable plumage variation across its extensive range.)

In New Jersey it is a common migrant and common breeding bird from late April to mid-September. It usually breeds in wet habitats, with the female constructing the nest in a few days (the breeding period is from mid-May to mid-June). The Yellow Warbler is frequently parasitized by the Cowbird, although the warbler often rebuilds the nests over its own and the Cowbird's eggs.

CHESTNUT-SIDED WARBLER *Dendroica pensylvanica* (Linnaeus)

The Chestnut-sided Warbler breeds across southeastern Canada, and through northeastern states south into the Smoky Mountains. It winters in lower Central America (mainly Costa Rica). It is a New Jersey migrant in May and late August to mid-September. It also breeds in second-growth areas of northern New Jersey woodlands, and rarely on the coastal plain of southern New Jersey or in the southwestern portion of the state. Apparently it was quite rare before the last half of the nineteenth century, but then forests began to be extensively removed, secondary clearings became common, and the warbler increased with the increase in nesting habitat. (The single clutch is laid in early June.) Isolated breeding areas near the coast have been recorded in recent years (e.g. Manahawkin).

MAGNOLIA WARBLER *Dendroica magnolia* (Wilson)

The Magnolia Warbler breeds across Canada and south through New England into the mountains of Pennsylvania and West Virginia. In New Jersey it is a common migrant (May and September). It is also a local breeder (late May into July) in evergreen groves with thickets, at higher elevations in northern New Jersey (at least formerly; its current status is uncertain). The grassy nest is hidden in low branches, and only the female incubates.

A high count of 30 was obtained at Bound Brook on September 21, 1976.

CAPE MAY WARBLER *Dendroica tigrina* (Gmelin)

The Cape May Warbler breeds in spruce forests across Canada, south into New England. Most individuals winter in the West Indies. In our area it is a migrant, usually uncommon in spring (May) and fairly common in fall (mid-September to early October). Its numbers vary from year to year. (During migration individuals often pause for several days to take up temporary residence at a favored evergreen, grapevine, or the like.)

High counts of 99 and 132 were obtained at the Island Beach* bird-banding operation on September 19, 1976, and May 28, 1980, respectively.

BLACK-THROATED BLUE WARBLER *Dendroica caerulescens* (Gmelin)

The Black-throated Blue Warbler breeds throughout southeastern Canada and the northeastern states, south to the Smoky Mountains. In the winter it is one of the more common migrant warblers in the West Indies.

In New Jersey it is a migrant (early May and late August) and a scarce breeding bird at scattered localities in northern counties. Our nesting areas include wooded stands with rhododendron thickets. The nest is built in June, usually in a shrub and close to the ground.

YELLOW-RUMPED or MYRTLE WARBLER

Dendroica coronata (Linnaeus)

This species is an abundant breeding bird in northern forests from Alaska, across Canada and south in northeastern states, to Pennsylvania. The winter range includes the southern United States (some along the Atlantic coast to New England), Mexico, and the West Indies.

In our area the Yellow-rumped Warbler is an abundant migrant (mid-April to mid-May and early October to mid-November). It is also a winter resident, mostly along the coast (where there are bayberry shrubs). Very rarely there are summer reports, and there is at least one documented New Jersey nesting in recent years, from the Pequannock Watershed* area in the summer of 1979. (It regularly nests as nearby as the Adirondacks.)

The western race of this species, the yellow-throated "Audubon's Warbler," has been reported as an accidental (e.g. at Bound Brook on May 10, 1970). There are a few records of the Audubon's Warbler from our neighboring states.

A high count of 4,000 Yellow-rumped Warblers was made at Higbee Beach,* Cape May.

BLACK-THROATED GRAY WARBLER

Dendroica nigrescens (Townsend)

The occurrence of this western species in eastern states is accidental, and most records have been from the coast. In New Jersey this very rare fall and winter visitor has been recorded as follows: Mickleton (April 21, 1957), Tuckerton (September 23, 1956), Barnegat (September 1, 1954), Island Beach* (September 29, 1962, and 1 banded in the fall of 1969), Franklin (November 7–10, 1964), Mountainside (November 16, 1969), Stone Harbor (September 27, 1964), Cape May (September 13, 1974, and September 30, 1961), and Princeton (late December 1982 into January 1983).

TOWNSEND'S WARBLER

Dendroica townsendi (Townsend)

The Townsend's Warbler nests in northwestern coniferous forests and comes to the eastern states only as an accidental visitor in winter. It is generally recorded at feeding stations (e.g. at suet).

The first New Jersey record came from Princeton where an adult male visited a feeder from December 23, 1971, to January 15, 1972. Other recent sightings come from Vincentown (adult male, December 5, 1977–February 21, 1978), Harmony (December 11–13, 1979), and Sandy Hook (December 16–18, 1980).

BLACK-THROATED GREEN WARBLER *Dendroica virens* (Gmelin)

The Black-throated Green Warbler breeds across southeastern Canada, through New England south to the Smoky Mountains, and along the coast of the Carolinas. Most individuals winter in Central America and the West Indies. In New Jersey it is a migrant (May and September) and a local breeder in evergreens (hemlock stands in northern New Jersey, and cedar swamps in the Pine Barrens*). The southern breeding locations include parts of Ocean, Burlington, and Atlantic counties (1980). Records of summer birds in the Pine Barrens* date back to at least 1935 (several in two cedar swamps of Ocean County). Generally the four-to-five egg clutch is laid in June. A high density of breeding birds, about 50 pair, was recorded at Weekstown in June 1980.

BLACKBURNIAN WARBLER *Dendroica fusca* (Müller)

The Blackburnian Warbler breeds across southeastern Canada, south into the Smoky Mountains. It winters from lower Central America into northwestern South America.

In New Jersey this warbler is a migrant (May and late August–September), and a local breeder at higher elevations in northern parts of the state. It nests in coniferous stands (e.g. hemlock and spruce), and in general it is very difficult for an observer to find the treetop nest. Nevertheless, there were numerous breeding reports in the summer of 1979.

YELLOW-THROATED WARBLER *Dendroica dominica* (Linnaeus)

The Yellow-throated Warbler breeds in the southeastern United States, north to New Jersey. It winters in Central America and the West Indies.

This southern species is rare in New Jersey, and it is seldom encountered except at its limited breeding areas. In southern parts it breeds near the Delaware Bay shore in Cumberland and Cape May counties (first nesting in 1922). In northern New Jersey it breeds at Raven Rock, by the Delaware River (since 1954). These two separate breeding populations represent extensions of a southern race *(D. d. dominica)* and the midwestern race *(D. d. albilora)*, respectively.

A high count of 7 nesting pair was made near Dividing Creek on the Delaware Bay shore in 1979.

PINE WARBLER *Dendroica pinus* (Wilson)

The Pine Warbler breeds in pine forests through much of the eastern United States, but it is certainly more common in southern states than in the Northeast. It winters in the south, north along the coast to New Jersey.

The Pine Warbler is a summer resident (late March–early October) of the Pine Barrens,* nesting in areas of mature pitch pine. In northern New Jersey it has bred rarely at high localities where there are suitable evergreens (usually ridge-top pitch pines). Elsewhere it is a scarce migrant (April and late September–October). Occasionally individuals remain in winter, and there are usually several reported on the state's Christmas counts each year (often at feeding stations).

A high count of 20 was made at Fort Lee on April 20, 1915.

PRAIRIE WARBLER *Dendroica discolor* (Vieillot)

The Prairie Warbler is a local breeding bird in most eastern states (currently it is having a modest range expansion to the north). It winters in the West Indies and Florida.

In New Jersey it is a migrant and summer resident (late April to August). As a breeding bird it is distributed most commonly in the Pine Barrens* where it is especially fond of areas with new pine and scrub oak growth that has resulted after a wildfire. In northern New Jersey it uncommonly nests in shrubby fields. (In Pennsylvania the species often uses Christmas tree farms as a nesting habitat.) The nesting season extends from mid-May into June, with the female doing all nest construction and incubation.

PALM WARBLER *Dendroica palmarum* (Gmelin)

The Palm Warbler nests in spruce forests and bogs across much of Canada. It winters in tropical America, northward to southern states (coastally to Massachusetts). In New Jersey it is a migrant (April and mid-September to late October). Small numbers of individuals overwinter on our southern coast, or rarely inland and further northward in mild periods. Spring migration is dominated by the brightly colored race, *D. p. hypochrysea,* the "Yellow Palm Warbler." The paler western race, *D. p. palmarum,* dominates the fall migration (and winter).

A high count of 54 Palm Warblers was obtained at Island Beach State Park* on October 4, 1963.

BAY-BREASTED WARBLER *Dendroica castanea* (Wilson)

The Bay-breasted Warbler breeds across much of southern Canada, south into our northern states (e.g. New York in the Adirondacks). It winters in tropical forests from Panama into South America.

In New Jersey the Bay-breasted Warbler is strictly a migrant (May and September), variable in numbers from year to year. The species is more commonly recorded in the northern half of the state.

A high count of 22 was recorded at Englewood on May 19, 1920.

BLACKPOLL WARBLER *Dendroica striata* (Forster)

The Blackpoll Warbler breeds from Alaska, across Canada, southeast to higher elevations in New England. In New Jersey it is one of our most common migrant warblers (May and mid-September to mid-October). It is notably "late" in migration, peaking a week or more later than most other species.

As the Blackpoll nests as nearby as the spruce and fir forests in the Catskills of New York State, it is a possible future nester for New Jersey. Migration stragglers are common, however (Wood 1979), and records through June should not be taken as an indication of breeding.

A high count of 78 Blackpoll Warblers was made at Island Beach State Park* on September 23, 1963.

CERULEAN WARBLER *Dendroica cerulea* (Wilson)

The Cerulean Warbler breeds in eastern North America and winters in northwestern South America. In New Jersey it is an uncommon summer resident (May–August) in the northern third of the state. Formerly it was quite rare; it entered New Jersey as a breeding bird along the Musconetcong River* and became widely recorded with nestings in the 1950s and 1960s. There have also been a few scattered breeding records from streamside woodlands in southern New Jersey, in the Pine Barrens.* The nest is built in May or June, in a tall deciduous tree (often oak or sycamore). Away from the local breeding areas the species is a very rare migrant (e.g. Institute Woods* at Princeton, and Hutcheson Memorial Forest* at East Millstone).

The first nesting in the state came from Millington in Somerset County, 1947. One of the most continuous breeding sites has been Raven Rock, in Hunterdon County, since 1955.

BLACK-AND-WHITE WARBLER *Mniotilta varia* (Linnaeus)

The Black-and-white Warbler breeds throughout most of the eastern United States and much of southern Canada. It winters in tropical America, north to our southern states. In our area it is a common migrant (mid-April to mid-May and September) and a summer resident. The species nests (mid-May to mid-June) in wooded areas throughout, but it is less common on the coastal plain. The ground nest is well hidden amid debris, and incubation of the single brood is by the female.

AMERICAN REDSTART *Setophaga ruticilla* (Linnaeus)

The Redstart breeds across much of North America and winters widely in tropical America, including the West Indies. It has been a dominant species of eastern deciduous forests, and it is a summer breeding bird throughout most of New Jersey. It will nest from early May into June, with the female doing all nest construction and incubation (nests are often in roadside trees).

Like so many other long-distance migrants the Redstart appears to have become less numerous in recent decades; formerly 50 or more might be seen during a migration wave in a single day, whereas today one is lucky to see 15–20.

PROTHONOTARY WARBLER

Prothonotaria citrea (Boddaert)

The Prothonotary Warbler breeds in much of the eastern United States, but it is most common in southern areas. It winters from Mexico into South America. In New Jersey this striking species is a rare migrant and local breeder (late April through August). It nests in tree cavities in wooded swamps, primarily in southern counties along the Delaware Bay shore. There are also numerous isolated nesting records from northern New Jersey including Great Swamp,* River Vale, Bound Brook, Bull's Island State Park,* and the Wallkill River.* Note that all of these are riparian sites. It has nested as far north as Sussex County (in 1979), and it has bred fairly frequently in parts of central New Jersey (e.g. Charles H. Rogers Refuge,* Princeton).

WORM-EATING WARBLER

Helmitheros vermivorus (Gmelin)

The Worm-eating Warbler breeds throughout the central areas of the eastern United States, northeastward into New England. It winters in Central America and the West Indies.

In New Jersey it is generally an uncommon and local summer resident (May to early September), nesting on the ground in forested areas of northern New Jersey and rarely in swampy woodlands near the Delaware Bay shore of southern New Jersey. The first suspected breeding in our southern parts came near Dividing Creek in 1977 and 1979. In 1980 2 territorial birds were found at Bear Swamp* in Cumberland County.

In many areas of its breeding range, both in New Jersey and elsewhere, the Worm-eating Warbler has shown significant population declines, especially where forests have become fragmented.

A high count of 9 came from Higbee Beach* on August 15, 1979.

SWAINSON'S WARBLER

Limnothlypis swainsonii (Audubon)

The Swainson's Warbler is a species of southeastern states, wintering in Mexico and the West Indies. In New Jersey it is a very rare migrant, occurring with "overflights" northward beyond its usual range in spring migration. As it breeds northward to Maryland (e.g. Pocomoke Swamp) it is perhaps surprising that we do not have more New Jersey records.

The first state record came from Linwood (May 1968) and the second from Island Beach State Park* (banded on May 17, 1979). (There were unconfirmed reports of a singing bird at a cedar swamp in southern New Jersey during a recent June.) Cape May had a recent Swainson's Warbler, at Higbee Beach* on May 12, 1981. (There are several recent records from Pennsylvania and New York, again in May.)

OVENBIRD *Seiurus aurocapillus* (Linnaeus)

The Ovenbird nests in deciduous or mixed forests across much of central and eastern North America. It winters in Central America and the West Indies, and rarely in the southern states. In New Jersey it is a common migrant and summer resident (late April to September), breeding in forested areas throughout the state. (In 1980, for example, it was reported as the most common and widely distributed bird of the cedar swamps in southern New Jersey.)

It was formerly more common and recent decades of declines may be related to forest fragmentation and high Cowbird parasitism. The single clutch is laid between mid-May and mid-June. The leaf-covered ground nest is notoriously difficult to find.

NORTHERN WATERTHRUSH *Seiurus noveboracensis* (Gmelin)

The Northern Waterthrush breeds across Canada and New England, south in mountains to West Virginia. It winters widely in wet habitats in the tropics (e.g. mangrove swamps).

In New Jersey the Northern Waterthrush is a widespread migrant (May and late August–late September). It is also a summer resident at higher areas of northern New Jersey, nesting in wooded swamps (e.g. 8 pair nesting in Kuser Bog* in 1979). There are nesting reports into central portions of the state (recently in Monmouth and Ocean counties), and singing birds in summer have been found even further south (e.g. Cumberland County, 1981).

A high count of 16 breeding birds was reported from the Pequannock Watershed* area in the summer of 1973.

LOUISIANA WATERTHRUSH *Seiurus motacilla* (Vieillot)

The Louisiana Waterthrush breeds throughout most of the eastern United States and winters in Central America and the West Indies. In New Jersey it is a scarce migrant (mid-April to early May and August to early September); it is rarer along the coast. It is a summer resident in wet woodlands in northern New Jersey, and rarely in southern swamps (e.g. near the Delaware Bay shore). High breeding densities come from such northern areas as the Pequannock Watershed* and Stokes State Forest* (along streams). The single clutch is laid between late April and early June. The first breeding in southern New Jersey came from Cape May in 1920. In 1979 and 1980 4–5 pair were nesting in Cumberland County.

KENTUCKY WARBLER *Oporornis formosus* (Wilson)

The Kentucky Warbler breeds throughout the eastern United States, north to New Jersey. It winters in tropical forests of Central America and northern South America. In the first half of this century it was a very rare migrant and breeder; now it is generally rare to uncommon as a summer resident (May–August) in deciduous woodlands. The nesting localities are scattered throughout much of the state, and eggs are laid between mid-May and late July.

Reports from the nineteenth century imply that the species was more widely distributed then, but the reason for its subsequent decline remains a mystery. (There were no nest records for New York State from 1942 to 1973, for example.)

CONNECTICUT WARBLER *Oporornis agilis* (Wilson)

The Connecticut Warbler breeds in central Canada and winters in South America. The spring migration route is interior (along the Mississippi Valley), whereas in fall most move south along the Atlantic coast.

In New Jersey the Connecticut Warbler is indeed a rare fall migrant (late September to mid-October), primarily at moist thickets in the northern half of the state, and along the coast. It is extremely rare in spring migration (May), but there have been records at this season (e.g. an adult male collected at Fort Lee on May 25, 1917).

High counts include 4 at Mickleton on September 26, 1964, and a remarkable 13 at the Island Beach State Park* banding program on September 28, 1963.

MOURNING WARBLER

Oporornis philadelphia (Wilson)

The Mourning Warbler breeds across south-central and eastern Canada, and in northeastern states south through the mountains of Pennsylvania. In New Jersey it is a scarce migrant (May and September), primarily north of the coastal plain. This thicket-inhabiting warbler is most readily found on spring migration when its loud calling reveals its presence. In view of nearby nesting in New York and Pennsylvania it is possible that the Mourning Warbler might sometime nest at higher elevations in northwestern New Jersey.

High counts include 4 at Princeton (Institute Woods*) during a day in late May 1978, 5 at this same location on May 30, 1983, and 8 at the Island Beach State Park* banding program on September 2, 1963.

MacGILLIVRAY'S WARBLER

Oporornis tolmiei (Townsend)

The MacGillivray's Warbler breeds in the western United States and is only an accidental visitor in the East. Identification is difficult as individuals of the Mourning Warbler can also have the "broken eye ring" used in the identification of most MacGillivray's. (In western Canada there is some hybridization, and the two species are sometimes considered conspecific.)

Better reports for New Jersey included 1 banded and measured at Troy Meadows* on September 12, 1976, and 1 netted and photographed at Island Beach* on May 28, 1979. (There have been records from other Atlantic Coast states such as Massachusetts and Connecticut.)

COMMON YELLOWTHROAT

Geothylpis trichas (Linnaeus)

The Yellowthroat breeds throughout most of North America. It winters in the American tropics and southern states (north to Long

Island, New York). It is a common summer resident throughout New Jersey, from May through September. Small numbers are seen later in the fall, but it becomes quite rare by winter. Only a few individuals are recorded in the state on each year's Christmas counts.

Yellowthroat nesting densities can be very high in favorable marshes (e.g. 17 nests in half an acre; Harrison 1975). Two broods are usually attempted between early May and August.

HOODED WARBLER *Wilsonia citrina* (Boddaert)

The Hooded Warbler breeds throughout most of the eastern United States and winters in Central America and the West Indies. In New Jersey it is a scarce migrant (May and August). Summer breeding birds are found in three distinct habitats: rich woodlands in northern New Jersey, cedar swamps in the Pine Barrens,* and red maple swamps near the Atlantic coastline. In southern New Jersey it is most common in the cedar swamps along the Delaware Bay shore in both Cumberland and Cape May counties (1980). Nesting dates extend from late May through July.

A high count of 14 was made in the area of Wawayanda State Park* on July 4, 1979.

WILSON'S WARBLER *Wilsonia pusilla* (Wilson)

The Wilson's Warbler breeds from Alaska south to the western states and across Canada south to New England. Most winter in Central America.

In New Jersey it is an uncommon migrant (May and September), often varying in abundance from year to year. Despite published suggestions to the contrary (e.g. Peterson 1980), there are no documented breeding records from New York or Pennsylvania.

A high count of 12 came from the Island Beach State Park* banding program on September 12, 1976.

CANADA WARBLER *Wilsonia canadensis* (Linnaeus)

The Canada Warbler breeds across much of southern Canada, south through the New England states and in mountains south to the Smoky Mountains. It is a common migrant (late May and late August to mid-September) in New Jersey. It is also a summer resident in the northern half of the state, nesting in wet woodlands. The grassy ground nest has three to five eggs in early June.

Nesting in more southern areas was established in 1980 at Assunpink Wildlife Management Area,* Monmouth County. In the same year territorial birds were also found at single locations in Cumberland, Ocean, and Atlantic counties.

YELLOW-BREASTED CHAT *Icteria virens* (Linnaeus)

The Yellow-breasted Chat breeds throughout almost all of the United States, albeit rarely in many areas. Most winter in Central America. In New Jersey it is a rare migrant and local summer resident, nesting in scrubby fields. The nesting period extends from late May to mid-July, and the single clutch is incubated solely by the female. A few individuals are sometimes reported in New Jersey during the winter (especially December), usually at feeding stations.

TANAGERS / *Thraupinae*

SUMMER TANAGER *Piranga rubra* (Linnaeus)

The Summer Tanager breeds in southeastern states, north and west to Iowa, Ohio, and New Jersey. It winters widely in Central America and the West Indies, south into South America. In our region this southern species is a rare migrant in spring. (Individuals occasionally "overshoot" their destination and migrate further north than normal.) It is also a scarce local breeder (late May to mid-July) in a few localities of southern New Jersey, but possibly it is increasing. During the mid-nineteenth century the Summer Tanager was recorded nesting in several southern New Jersey localities, and it may have been more numerous than it is today.

Recent nesting has been noted at Dividing Creek, Cumberland County, in 1975 and subsequent summers with 5 pair or more (the only previous New Jersey nesting in this century was in 1955). Several pair were breeding in the northern Pine Barrens* during 1980. It favors open areas of mixed pine and oak forest, with nests and eggs very similar to those of the Scarlet Tanager.

SCARLET TANAGER *Piranga olivacea* (Gmelin)

The Scarlet Tanager breeds in the northern half of the eastern United States and just into southern Canada. It winters in forests of northwestern South America. In our area it is a common migrant and widespread breeder (present from May to September). Although it nests primarily in the deciduous woodlands of north and central New Jersey there are breeding records (mid-May to August) throughout the state. The frail nest is usually located on a high limb, often in an oak.

High counts include 14 pair nesting at the Greenbrook Sanctuary* (summer of 1979), and 60 at Ramsey during migration (May 15, 1947).

WESTERN TANAGER *Piranga ludoviciana* (Wilson)

This colorful tanager breeds in the western United States and Canada. It is generally a very rare visitor in eastern states, but it may be regular in Florida during winter.

New Jersey has at least 14 records, starting with March 1938 at Island Beach* (a bird present for several days). Most reports come within the period of November to March, with December being most frequent for sightings. Sightings are widely scattered, both inland and along the coast (e.g. Maplewood, Brigantine National Wildlife Refuge,* Myersville, and Park Ridge). At least 2 Western Tanagers have been banded in New Jersey: 1 at Montclair on November 10, 1956, and 1 at Braddock's Mill Lake* from December 30, 1972, into January 1973. A well-documented bird was seen by about 200 observers at Franklin, Sussex County (January to April 1953). A summer record was obtained at Riverdale on August 21, 1962. More recent sightings include: 1 at Cape May (December 16, 1979–January 10, 1980), a male at Middletown in northern Cape May County (May 16–17, 1981), and 1 near Kendall Park (December 20, 1981, into January 1982).

GROSBEAKS AND BUNTINGS
Cardinalinae

NORTHERN or COMMON CARDINAL
Cardinalis cardinalis (Linnaeus)

The Cardinal is a resident throughout the eastern United States. For several decades it has been gradually expanding its range to the north and west. In our state it is a common permanent resident in woodland ecotones or shrubby suburbs throughout. Formerly this southern species was scarce in northern New Jersey, but it has made dramatic increases in abundance and range since 1940. For example, the Sussex Christmas count tallied 35 in the early 1950s, but by 1961 it was up to 190 (it has not increased since then).

Cardinals nest between mid-April and August, and often two broods are raised. The nest is usually low, but well hidden in thickets or small trees (including ornamental conifers in yards).

ROSE-BREASTED GROSBEAK
Pheucticus ludovicianus (Linnaeus)

The Rose-breasted Grosbeak nests in deciduous forests from west-central Canada southeast to New Jersey and, at higher elevations, to the Smoky Mountains. It winters widely in tropical America, south to Peru.

In our area it is a common migrant (May and September). It is also a summer resident in woodlands of northern New Jersey, uncommonly south to Burlington County. Very rarely, there are a few winter records, usually at feeding stations. The Rose-breasted Grosbeak nests in early June, with both sexes incubating (often an incubating male will sing).

A remarkable high count of 100 was made at Pohatcong Creek Valley* on September 19, 1976.

BLACK-HEADED GROSBEAK *Pheucticus melanocephalus* (Swainson)

The Black-headed Grosbeak breeds throughout the western United States. It is a very rare visitor to northeastern states.

Most New Jersey records are from the winter, when birds are at feeding stations for at least several days. Sightings include: Englewood (January–February 1955), Allendale (January–February 1957), Bridgeton (January–March 1957), Maplewood (February–April 1960), Cape May (Christmas count, 1960), Asbury Park (January 1965), Wenonah (February 1965), Franklin (December 1966), Woodbury (December 1971), Princeton (December 1972), and Rumson (December 1974).

BLUE GROSBEAK *Guiraca caerulea* (Linnaeus)

The Blue Grosbeak nests across the southern half of the United States, northward along the Atlantic coast to New Jersey. It winters in Central America and the West Indies. (There are also breeding populations in tropical America.)

In our area this grosbeak was formerly a very rare migrant, most frequently reported in fall near the shore. It appears to be established as a breeding species in south-central New Jersey (e.g. in Mercer and Monmouth counties).

The first New Jersey nesting was near Camden in 1829 (Audubon). By the 1950s there were records from areas of Cumberland County, and by the end of the decade successful nesting was noted at Middlebush, Somerset County. Other breeding localities have included Metuchen in Middlesex County and Princeton in Mercer County, as well as more northerly spots. Recently a substantial colony (8 pair in 1977) was established at Assunpink Wildlife Management Area* in Monmouth County. The Blue Grosbeak nests through the summer months and can raise two broods.

INDIGO BUNTING *Passerina cyanea* (Linnaeus)

The Indigo Bunting breeds throughout the eastern United States and winters in Central America and the West Indies. In our area it is a migrant throughout. It is also a New Jersey summer resident (May through September) at woodland edges, primarily in the northern half of the state, although it has nested south through Cape May County. The breeding season extends from May to August, with one or two broods. Very rarely, individuals are recorded in northeastern states during the winter (e.g. 1 at a feeder in Trenton, December 23, 1980).

A high count of 69 was made at Blairstown in the summer of 1966.

PAINTED BUNTING *Passerina ciris* (Linnaeus)

The Painted Bunting nests in southern states, north to the coast of North Carolina. It is probably a very rare visitor to our region, but records of this species, especially older records, are difficult to evaluate as it has often been kept as a cage bird (now illegal). Recent sightings, however, should involve wild birds in at least some if not all cases. More recent records include 1 (adult male) at Cape May Point (May 4–5, 1958), 1 at a feeder at Lebanon (December 1976–January 1977), and 1 banded and photographed at Island Beach State Park* (May 12, 1980).

DICKCISSEL *Spiza americana* (Gmelin)

The Dickcissel is primarily a breeding bird of agricultural fields in the Midwest. It winters from Mexico to northern South America, again often in agricultural areas.

In New Jersey it is a rare migrant (mostly in the fall and usually heard flying overhead, giving a buzzy call). It is also a rare winter visitor to feeders, and has been a very local, very rare, breeding species in recent years. This species formerly had a fairly widespread breeding population in the mid-Atlantic states, including the northern counties in New Jersey. Nesting in eastern states had essentially ended by the 1860s. (The last New Jersey nesting in the nineteenth century was in 1880.) There was an isolated nesting pair at Plainfield in July 1904, but recent breeding did not occur until 1973 near Atlantic City and 1974 at the Hackensack Meadows.* (The ground nests are very difficult to find.)

A high count of 5 migrants (banded) was made at Island Beach State Park* on October 6, 1962.

TOWHEES, SPARROWS, AND LONGSPURS
Emberizinae

GREEN-TAILED TOWHEE

Pipilo chlorurus (Audubon)

The Green-tailed Towhee is a bird of the western states; it occurs in the East only as an accidental visitor. In New Jersey it is usually recorded at feeding stations in winter, and the birds may be resident for weeks or months.

Sightings are mostly from the northern part of the state, with examples from Bergen County (December 1939–January 1940), Whitesville (December 1962), Newton (January–April 1961), and Flemington (November–December 1968). The first record for southern New Jersey came from a feeder at Mickleton (November 20, 1975–April 10, 1976).

RUFOUS-SIDED TOWHEE

Pipilo erythrophthalmus (Linnaeus)

The Rufous-sided Towhee breeds throughout much of the United States, but in winter it retreats southward from northern states. In New Jersey it is a common summer resident (April through October) with small numbers remaining through the winter months. As a breeding bird it is quite abundant in the Pine Barrens,* often outnumbering all other local species. Winter sightings from northern New Jersey often come from feeding stations.

Favored nesting habitats include brushy fields and woodland thickets. This ground nester typically has two broods between late April and mid-August.

The western race, the "Spotted Towhee" *[P. e. montanus]* has been seen in New Jersey on a few occasions (e.g. Metuchen on December 24, 1952, and near Englewood in late December 1967).

BACHMAN'S SPARROW *Aimophila aestivalis* (Lichtenstein)

The Bachman's Sparrow is a breeding bird of the East, primarily in southern states. It winters from Texas to North Carolina. It is strictly an accidental visitor in New Jersey, and there are only two state records, although the species breeds as nearby as Maryland and it has bred in Pennsylvania (four counties). Our reports are of a singing male collected at Fort Lee on May 9, 1918, and another singing bird seen by two observers at Atsion, Burlington County, on June 16, 1957.

CASSIN'S SPARROW *Aimophila cassinii* (Woodhouse)

The Cassin's Sparrow has a restricted breeding range in the western United States. It is only an accidental visitor to eastern North America, and was once recorded in New Jersey. One was caught while birds were being banded at Island Beach State Park* during fall migration on September 22, 1961.

AMERICAN TREE SPARROW *Spizella arborea* (Wilson)

The Tree Sparrow breeds in arctic thickets of Alaska and northern Canada. In winter it occurs across most of the northern United States, south on the Atlantic coast to Virginia. In our state it is a common winter resident (November through March) in brushy fields. Usually it occurs in small flocks, often with a few Field Sparrows.

A high count of 998 was made on the Walnut Valley Christmas count in 1974–1975.

CHIPPING SPARROW *Spizella passerina* (Bechstein)

The Chipping Sparrow breeds throughout much of North America, and it winters from Mexico and southern states northward along the Atlantic coast (rarely to Rhode Island). It is a common New Jersey migrant and summer resident (April–October) in open areas. The Chipping Sparrow is fond of lawns and will nest in many suburban situations. It is generally rare in winter, but small numbers do remain.

It will have two broods between early May and late August. The nest is often low, in a short tree or shrub, and it is characteristically lined with hair.

CLAY-COLORED SPARROW *Spizella pallida* (Swainson)

As a breeding bird the Clay-colored Sparrow is primarily a midwestern species of Canada and the northern United States. It has, however, had an eastward range expansion that now includes parts of Ontario and even New York State (e.g. 1960, 1970, 1974, and 1975).

In New Jersey it is a rare visitor, usually along the coast in fall migration. It is now recorded annually (formerly it was accidental in our state).

Earlier records include Beach Haven (May 8, 1932), Barnegat Light (September 12, 1953), and Cape May Point (October 10, 1953). It has been seen annually since 1954, with three records in 1957, at least five in 1958, and so on. Most sightings are in September and October, but there have recently been summer birds (e.g. several singing near Morristown in June and July, 1978), and even overwintering ones (e.g. at a feeder in Boonton, January–April 1978).

FIELD SPARROW *Spizella pusilla* (Wilson)

The Field Sparrow breeds and winters in the eastern United States. In New Jersey it is primarily a summer resident (April into October) in old fields and pastures. The Field Sparrow will nest from May through August, with two broods on the average. Multiflora rosebushes are a favored support for the grassy nest.

The Field Sparrow is also a winter resident, especially in southern and coastal portions of the state. The Cape May Christmas count tallied 303 Field Sparrows on December 20, 1981, for a seasonal high count.

VESPER SPARROW *Pooecetes gramineus* (Gmelin)

The Vesper Sparrow breeds in fields across most of central North America. It winters in tropical America and southern states, north along the coast to Long Island, New York. It is a New Jersey migrant (April and October), and formerly a widespread summer resident, nesting in pastures and agricultural fields. As a breeding species in New Jersey it has very much declined, and nesting sites are now local and rare (mostly in Warren and Somerset counties, 1981). It is a scarce winter resident, primarily in our southern coastal counties.

LARK SPARROW *Chondestes grammacus* (Say)

The Lark Sparrow breeds throughout much of the western and central United States, east to Ohio. It winters from southern states into Central America. There are a few older breeding records from some East Coast states (e.g. Maryland and Virginia). In New Jersey it is a rare visitor, and most records are from the coast during fall migration (August through mid-October). A few are now reported each year in New Jersey.

Earlier records include the first New Jersey specimen (November 26, 1885), from Schraalenburg (?), and sightings at Seven Mile Beach* (August 21, 1927, and August 15, 1937), Cape May (May 12, 1935), Brigantine (August 15, 1935), Union County (October 28, 1928), and Essex County (September 19, 1934). There are winter records (e.g. Bound Brook on January 4, 1959, and Ridgewood from January 14 to February 4, 1953). More recently 2 were at Princeton (December 14–21, 1980).

Recent high counts include 4 at Island Beach State Park* during a week in mid-September 1975, and 11 along the coast during the fall of 1979 (the last on October 28).

BLACK-THROATED SPARROW *Amphispiza bilineata* (Cassin)

The Black-throated Sparrow is a desert species of the southwest, east to Texas. In the Northeast it is only an accidental visitor. The first New Jersey record was of a bird at a feeding station near New Brunswick, from October 30, 1961, to April 23, 1962; it was seen by many and photographed. Another was recorded at the North Arlington cemetery from December 14, 1974, to January 19, 1975.

LARK BUNTING *Calamospiza melanocorys* (Stejneger)

The Lark Bunting nests in prairies of central North America, and it strays to the Atlantic Coast states only as an accidental visitor. We have a few records from our state, which include 1 at Cape May Point (September 16, 1956), 1 at the Delaware Bay shore (May 26, 1962), and a male at Higbee Beach* (May 26, 1969). Also, the species has been banded at Island Beach State Park* during migration.

SAVANNAH SPARROW *Passerculus sandwichensis* (Gmelin)

The Savannah Sparrow breeds throughout most of North America, north of the central United States. It winters along the coasts and in southern states, into the tropics. In New Jersey it is a migrant (mid-March through mid-April and October), and a scarce local breeder, primarily on short-grass fields or salt marshes. It also winters in southern sections of the state, especially near the coast. Recent nesting areas included rural sections of Warren County. Formerly it was more common as a breeding bird (e.g. 125 nesting pair at Newark Meadows* in 1936). The grassy ground nest is usually constructed in May, although a second brood may be possible in late June.

The large pale race, the "Ipswich Sparrow," was formerly considered a distinct species. It is a scarce local winter resident from the first week of November, principally along dunes on Ocean County's coast (e.g. Holgate, Brigantine Island,* and Longport). A high count of 25 "Ipswich Sparrows" was made at the south end of Brigantine Island* early in 1976.

GRASSHOPPER SPARROW *Ammodramus savannarum* (Gmelin)

The Grasshopper Sparrow breeds across much of the United States, northward into Canada. (There are also breeding populations in tropical America, south to Ecuador.) North American birds winter in southern states, Central America, and the West Indies. There have been serious population fluctuations and declines in many nesting areas of the Northeast.

In New Jersey it is now an uncommon and local summer resident (late April through October), in old fields. It can raise two broods between mid-May and mid-August. Formerly it was more widespread, and some of its decline is probably related to habitat loss. Very rarely, individuals are reported in New Jersey during winter months (usually there are none reported from Christmas counts).

High counts at breeding colonies include 19 pair at the Newark Meadows* in 1935, and 30 pair near Atlantic City in 1981. A high tally of banding came from Island Beach State Park* with 9 on October 6, 1962.

HENSLOW'S SPARROW

Ammodramus henslowii (Audubon)

The Henslow's Sparrow nests in weedy fields in the northeastern United States west to Kansas and south through Virginia. Its breeding range has been significantly reduced in many areas.

In New Jersey it has been a rare and local summer resident, breeding in brushy fields, marsh edges, and meadows. The nesting period extends from mid-May through July. Formerly it was more widely distributed and it was reported as common near some salt marshes (e.g. in the 1930s). Six pair nested near New Brunswick in 1952. In 1979 and 1980 there were no known New Jersey nestings. (A few summer birds have been documented since then.)

LE CONTE'S SPARROW

Ammodramus leconteii (Audubon)

The Le Conte's Sparrow breeds in marshes of central Canada and the northern United States, east to Lake Erie. It migrates to southeastern states for the winter (north to South Carolina). In our region it is a very rare visitor. There are apparently two well-documented records from our state. The first was of a singing bird well studied by several observers at Jamesburg, May–June 1964. The second was a bird banded at Island Beach State Park* on October 12, 1968. Sight records for this secretive and sometimes difficult to identify species should be made very carefully. There are a few published reports from Pennsylvania (2 in fall) and New York (2 in fall, 2 in spring).

SHARP-TAILED SPARROW

Ammodramus caudacutus (Gmelin)

The Sharp-tailed Sparrow has a fragmented breeding range; the eastern population breeds coastally from Nova Scotia to North Carolina. In New Jersey it is a common summer resident (May through September) in salt marshes. Sharp-tailed Sparrows can have two broods between mid-May and mid-August. As a migrant it is common, especially in the fall. Small numbers of individuals remain in winter months. Inland reports are rare, during either spring or fall migration.

Several distinctive forms, representing various subspecies plumages, are found in New Jersey during an annual cycle. (The three most distinctive types are illustrated in Peterson's [1980] *A field guide to the birds*.)

A high count of 100 Sharp-tailed Sparrows was made near Villas, Cape May County, on September 3, 1925.

SEASIDE SPARROW *Ammodramus maritimus* (Wilson)

The Seaside Sparrow is a coastal resident along most of the Atlantic and Gulf coasts of the United States. In New Jersey it is primarily a summer resident in salt marshes (April through October), breeding along both the coast and the Delaware Bay shore. The nest is often built in a small shrub (e.g. marsh elder), and the single clutch is laid in June. Formerly the species was more abundant; salt-marsh drainage and fill-in have reduced its nesting habitat. Small numbers are present along the coast in winter, and the state's Christmas counts usually total several dozen.

FOX SPARROW *Passerella iliaca* (Merrem)

The Fox Sparrow breeds across Canada and Alaska and in many western states. In the eastern United States it winters northward through New Jersey and along the coast, occasionally to Maine.

March and November are the migration months for Fox Sparrows passing through New Jersey, and often the passage is fairly brief, especially in the spring (one or two weeks). Some individuals remain into winter, especially at feeders, but it is usually uncommon at that season.

A truly remarkable high count of 700 was made at Franklin Lakes on March 25, 1933.

SONG SPARROW *Melospiza melodia* (Wilson)

The Song Sparrow has a broad breeding range across North America, and there is considerable geographic variation in plumage and even size. In New Jersey it is a permanent resident throughout, although it is less common in winter. The species is particularly numerous in shrubby areas, either in old fields or suburban plots. The nest is usually well hidden in ground vegetation, and two broods are often attempted between late April and the end of August. The species is one of the most frequent hosts of Cowbird eggs. A high of 526 Song Sparrows was tallied on the Boonton Christmas count of December 27, 1981.

LINCOLN'S SPARROW *Melospiza lincolnii* (Audubon)

The Lincoln's Sparrow breeds in brushy wet habitats of Alaska, Canada, the western United States, and northern New England (south to the Adirondacks of New York). It winters in southern states and Central America.

In New Jersey it is an uncommon migrant (May and early October). This secretive species is most often found in wet thickets or old fields. Very rarely, the species remains into winter, and it is recorded only as an accidental on the state's Christmas counts.

A high count of 15 was obtained at Somerset on October 11, 1979.

SWAMP SPARROW *Melospiza georgiana* (Latham)

The Swamp Sparrow breeds in freshwater habitats of Canada and the northeastern United States. It winters throughout most of the eastern United States. In New Jersey it is most common as a migrant. It is also a local summer resident (April through October), nesting in marshes and wet habitats throughout. (Frequently Swamp Sparrows nest in small, semicolonial groups.) Some individuals remain in winter, particularly along the coast.

A high count of 120 nesting pair was obtained at Troy Meadows* in the summer of 1947.

WHITE-THROATED SPARROW

Zonotrichia albicollis (Gmelin)

The White-throated Sparrow breeds in mixed forests through much of Canada and New England, south into the mountains of Pennsylvania and rarely West Virginia. In New Jersey it is a very common migrant and a common winter resident (October through early May). Very rarely birds remain into summer, and even nesting is reported in extreme northern New Jersey. The female builds the nest of grass and rootlets, usually on the ground. The first state nesting records came in 1971 from coniferous forests in the Pequannock Watershed.* Breeding pairs have been found in some other years (e.g. 1973 and 1980). The species regularly nests in nearby parts of New York State at higher elevations.

A Christmas count high of 752 was tallied at Cape May on December 20, 1981.

GOLDEN-CROWNED SPARROW

Zonotrichia atricapilla (Gmelin)

The Golden-crowned Sparrow is a bird of Pacific Coast states and British Columbia. It occurs as an accidental in northeastern states (e.g. Massachusetts). There have been several New Jersey records. The first was a bird at Cape May on October 7, 1962: it was seen by over 500 observers (!) because of the New Jersey Audubon meeting at Cape May. At a Boonton feeding station 1 individual was also seen by many observers, in several successive winters (including January until April 1976. One was seen at Warren County in early 1983.

WHITE-CROWNED SPARROW

Zonotrichia leucophrys (Forster)

In western North America the White-crowned Sparrow is a common and widespread breeding bird. In the East it breeds across northern Canada, and it winters in southern portions of the United States, northward into New Jersey. In our state it is mostly an uncommon migrant (late April into early May and October), and a scarce winter resident. There were very few winter records before the 1950s; recent milder winters and feeders may have contributed to the present frequency of overwintering birds.

High counts include 42 at the Red Bank airport (December 31,

1977), 46 on the Assunpink* Christmas count (December 27, 1980), and 86 at Mullica Hill (April 1965). One hundred and twenty were banded at Mullica Hill in the spring of 1965, and 234 were banded in the fall of 1966 at Island Beach State Park.*

HARRIS' SPARROW *Zonotrichia querula* (Nuttall)

The Harris' Sparrow breeds at the edge of the boreal forest in north central Canada and winters in midwestern states. In our area it is a very rare visitor, with most records from feeding stations during winter or fall migration. The first New Jersey record came from Morris County, on May 7, 1935. Subsequent sightings include: Mahwah (April 30, 1966), Stony Ford Sanctuary at Princeton (May 1966), Island Beach* (October 7, 1967), Sparta (September 21, 1968), near Princeton Junction (November 1968 into January 1969), Secaucus (December 26, 1971), and Mickleton (January 1975).

DARK-EYED JUNCO *Junco hyemalis* (Linnaeus)

The Junco is a breeding bird of coniferous forests, mostly in Canada and Alaska. It breeds south through New York State and at higher elevations to the Smoky Mountains. In New Jersey it is a common migrant and winter resident (October through April). Also, there are a few breeding reports from northern New Jersey. (The nest is usually built on the ground, amid tree roots or logs.) This species now includes the brown-sided race known as the "Oregon Junco," which is a rare winter visitor, usually at feeding stations. In addition, the large "White-winged Junco," a distinctive race from South Dakota, has been recorded at least once (a bird trapped and photographed with bird banding at Hopewell on January 8, 1976).

Breeding of the Dark-eyed Junco in the state came as early as 1949 at Stokes State Forest* (July 2), and later at the Ramapo Mountains* (July 26, 1953).

Individual Christmas counts in northern New Jersey often tally 1,000 or more Dark-eyed Juncos.

MCCOWN'S LONGSPUR *Calcarius mccownii* (Lawrence)

The McCown's Longspur is a species of western states. *New Jersey Nature News* published a report of 2 McCown's Longspurs on the Sussex County Christmas count, December 26, 1960. The report lacked critical details and has been considered insufficient. Confirmation is clearly needed for this hypothetical species in New Jersey. (The species has been recorded in Massachusetts.)

LAPLAND LONGSPUR *Calcarius lapponicus* (Linnaeus)

The Lapland Longspur breeds in the high arctic tundra of Alaska and Canada. It winters in many areas of the United States, albeit uncommonly. In New Jersey it is a rare winter visitor (December through February), usually within flocks of Horned Larks or Snow Buntings, on open fields. A high count of 25 came from Holgate on March 4, 1973 (the males in this flock were coming into breeding plumage).

CHESTNUT-COLLARED LONGSPUR *Calcarius ornatus* (Townsend)

The Chestnut-collared Longspur breeds in prairie areas of midwestern states and Canada. In East Coast states it is only an accidental visitor, and there now are only two records for New Jersey. A male in breeding plumage was found at Cape May on the remarkable date of June 18, 1980. It was seen by numerous observers and well photographed (*Records of N.J. Birds*, 1980, 6 [4]:74). New York State also had a recent coastal record (June 6–8, 1982, in Nassau County). In early 1984 one frequented Sandy Hook State Park.

SNOW BUNTING *Plectrophenax nivalis* (Linnaeus)

The Snow Bunting breeds in the high arctic tundra from Alaska eastward to Greenland and Labrador. In winter it regularly moves into southern Canada and the northern United States (south along the Atlantic coast through the Delmarva Peninsula).

In New Jersey this bunting is an erratic winter visitor (November through February), and it is usually uncommon. Perhaps it is most regular along the coast where flocks frequent dunes and beaches.

A high count of 150 was made at Boonton on January 26, 1930.

ORIOLES AND BLACKBIRDS
Icterinae

BOBOLINK *Dolichonyx oryzivorus* (Linnaeus)

The Bobolink breeds in fields across the northern United States and southern Canada. It winters in southern South America (e.g. the pampas of Argentina). In New Jersey it is a common fall migrant (late August to mid-September), especially along the coast (including the Delaware Bay shore). It is uncommon as a spring migrant throughout. It is now quite scarce as a summer resident with scattered breeding localities in fields, mostly in northwestern New Jersey farmland (especially Hunterdon County and northwestern Somerset County). Sometimes it nests south to Cumberland County. Formerly the Bobolink was a much more common nesting bird in our state, but the species has undergone serious population declines in the East. (The Red-winged Blackbird has had a concomitant increase in upland fields.) The Bobolink's nest usually contains eggs in mid-June, and early mowing can be dangerous to the young.

A high breeding density was recorded at Newark Meadows* in 1935 (93 pair), and more recently (1959) at farmlands of Hunterdon County.

RED-WINGED BLACKBIRD *Agelaius phoeniceus* (Linnaeus)

The Red-winged Blackbird is a common and widespread species, breeding throughout most of North America in marshes and fields. It winters in most of the United States, but is particularly numerous in the South and into Mexico. In New Jersey it is a common to abundant resident. In winter it is primarily distributed in our southern counties and along the coast, at winter roosts. With spring and summer it breeds throughout, particularly in marshes but increasingly in upland fields as well. Breeding extends from May into August, often with two broods. Males are polygamous, and females thus do all the incubation.

High counts come from Salem where December estimates can be 500,000 to several million.

EASTERN MEADOWLARK

Sturnella magna (Linnaeus)

The Eastern Meadowlark breeds in fields throughout most of the eastern United States. In the South its breeding range has expanded westward into Arizona.

In New Jersey the meadowlark is essentially a permanent resident in southern and coastal areas, and a summer resident (late March to October) in the remainder of the state. As with many grassland species it was formerly more common as a breeding bird. This polygamous species usually has two broods between mid-May and mid-August. The grass nest typically has a complete roof or canopy, and entrance is made by an opening on the side.

WESTERN MEADOWLARK

Sturnella neglecta (Audubon)

The Western Meadowlark primarily breeds in western states, but some nest as far east as New York State. In recent years there has been an expansion of breeding range. Although it is still a very rare visitor to New Jersey we can expect to see more frequent records in the future.

Our state's earlier sightings include: South Plainfield (April 28, 1940), Columbus (June–August 1951), Troy Meadows* in Morris County (May 2–9, 1938), Hopewell (May 1–10, 1953), and Bound Brook (May 15, 1954, and April 10, 1956). In the 1960s birds were noted at Blawenburg (July 4, 1961), at Plainsboro (June 30, 1964), and near Wrightstown (June 22, 1964). In almost all cases the birds were first detected by their distinctive call which is very different from that of the Eastern Meadowlark.

YELLOW-HEADED BLACKBIRD

Xanthocephalus xanthocephalus (Bonaparte)

The Yellow-headed Blackbird is a marsh bird of central and western North America, but as with the Western Meadowlark this species has been expanding its breeding range to the east (e.g. southeastern Ontario), and it may be expected more frequently in the future. Still, it is a very rare visitor to New Jersey.

Earlier records include an 1890 report from Ocean County, one in Camden County on September 1, 1917, and a bird at Cape May in mid-August 1923. Sightings come from all months of the year, but especially in fall and winter. Most are recorded in coastal areas (e.g. South Amboy, Forked River, and Ship Bottom), but regularly there are inland reports as well (e.g. Dutch Neck, Lafayette, and the Great Swamp*). Since 1960 the species has been recorded almost annually in the state, and in recent years there have been numerous records (e.g. 6 banded at Ship Bottom during August and September 1977). Occasionally the species visits feeders (e.g. at Rumson on September 5, 1979, and a male and female at Forked River in September–October 1979).

RUSTY BLACKBIRD *Euphagus carolinus* (Müller)

The Rusty Blackbird breeds in wet wooded areas across most of Alaska and Canada, southeast into New England. It winters throughout the eastern United States, but is most numerous toward the south. In New Jersey it is an uncommon migrant in wet forests and swamps (late March through April and mid-October through mid-November). Small numbers remain in winter where open ground remains unfrozen (e.g. at marsh springs).

High counts include 300 at Troy Meadows* on March 21, 1943, and 400 at the Raritan Arsenal* on December 23, 1939.

BREWER'S BLACKBIRD *Euphagus cyanocephalus* (Wagler)

The Brewer's Blackbird is a common bird in agricultural lands throughout much of the western United States. Its range is expanding eastward, and it now breeds as far east as Ontario. (It regularly winters east to the Atlantic coast in southern states.) Still, it is only an accidental visitor in New Jersey. State reports include 4 at Cape May on December 27, 1959 (and subsequently), Hampton Township on December 9–10, 1960, Wenonah on December 21, 1965, Livingston on December 5–6, 1970, and East Brunswick on November 11–23, 1978 (several observers; photographed).

BOAT-TAILED GRACKLE *Quiscalus major* (Vieillot)

The Boat-tailed Grackle is a resident in Florida, on the Gulf coast, and on the Atlantic coast north to New Jersey. As a local resident on the Jersey shore it has been expanding northward in recent years and has clearly become more numerous. It is nesting as far north as Island Beach* and may soon be breeding near Sandy Hook.

The first New Jersey records were in 1892, the next in 1936. By 1952 there were small breeding colonies discovered at Gandys Beach, Cumberland County (15 nests in three colonies on the Delaware Bay shore). By 1958 there were at least 75 pair nesting along the Delaware Bay. The bulky nest is built in marsh vegetation or in short trees. Our populations probably have a single brood, in May or June.

Most sightings are along the coast, but occasionally there are inland reports (e.g. 2 at Assunpink* in September 1979).

High counts included 97 at Cape May (January 9, 1974), 400 at Brigantine Island* (November 5, 1977), and 666 on the Cape May Christmas count (1981–1982).

COMMON GRACKLE *Quiscalus quiscula* (Linnaeus)

The Common Grackle breeds throughout much of central and eastern North America. In winter it retreats to southern states, although small numbers remain as far north as southern Canada. In New Jersey this grackle is an abundant migrant and common breeder throughout. It is locally common in winter at roosts. (Most grackles seen in our state are of the "bronzed" race, *Q. q. versicolor.*)

The Common Grackle typically nests in loose colonies, in a wide variety of habitats (e.g. forest, marsh, or suburban yard). A single brood is raised between April and July.

High counts include 70,000 at the Raritan Arsenal* in December 1952, and 500,000 at Salem in December 1979.

BROWN-HEADED COWBIRD *Molothrus ater* (Boddaert)

The Brown-headed Cowbird breeds throughout much of North America, and it is apparently increasing. It winters across the southern United States, with some remaining in the northern states in mild winters. In New Jersey the Cowbird is a common to abundant migrant, and a widespread breeder. It is generally uncommon in winter except at local roosts (where it may be numerous). It lays its eggs between April and the end of July; over 100 other species serve as effective hosts for this nest parasite! (Favored victims include the Yellow Warbler and the Song Sparrow.) Many species, including such forest birds as the Wood Thrush and the Ovenbird, may be suffering population declines related to the current high levels of Cowbird parasitism.

High counts include 35,000 at the Raritan Arsenal* on December 23, 1946, and 10,000 at Salem on December 29, 1979.

ORCHARD ORIOLE *Icterus spurius* (Linnaeus)

The Orchard Oriole breeds throughout most of the eastern United States. It winters from Mexico into northern South America (it is often seen in flocks in the wintering grounds).

In New Jersey this oriole is an uncommon migrant and local breeder (May–July). Generally it is more common in central and southern areas of the state, but even there it is often scarce. (In recent years populations have been high.) Orchards are a favored nesting habitat, but a wide variety of shade trees can be used. It usually has a single brood in June (the incubating female will be fed at the nest by her mate). Apparently the species was more common in the first half of this century.

A high count of 15 nesting pair was made at Assunpink* in June 1981.

NORTHERN ORIOLE *Icterus galbula* (Linnaeus)

The Northern Oriole breeds throughout most of the United States and parts of southern Canada. It winters in Central America and northwestern South America. In New Jersey it is a common migrant and widespread breeding bird (May–September), but scarce in the Pine Barrens.* Formerly it was very rare in winter, but recently small numbers appear to be staying at feeding stations (e.g. 4 at one feeder in Princeton). The species has a strong nesting preference for elms, and it might thus be indirectly affected by the Dutch elm disease. The distinctive woven nest has an opening at the top. A single clutch is laid in late May or early June.

Our eastern race is the "Baltimore Oriole," *I. g. galbula*. The western race, "Bullock's Oriole," *I. g. bullockii*, is a very rare visitor in fall and winter. Records of Bullock's include Cape May Court House (December 1953), Boonton (February–March 1953), Cape May (December 1960 and September 1966), Ridgewood (December 1964), Wood Ridge (October 7, 1969), and Piscataway (January 9, 1972).

A high count of 13 Northern Orioles was obtained at Hutcheson Memorial Forest,* Somerset County on July 20, 1967 (all banded here on that date).

CROSSBILLS AND FINCHES *Fringillidae*

BRAMBLING *Fringilla montifringilla* (Linnaeus)

The Brambling may occur in our region naturally as an accidental visitor. Records of this European species are questionable, however, as it is frequently a cage bird with reported escapes (e.g. Pennsylvania and New York records have been considered "apparent escapes").

The first New Jersey record was also a first occurrence for eastern North America. It was a male at Stanton, Hunterdon County, on December 17, 1958. A second record for our state came from Branchville in mid-April 1965.

PINE GROSBEAK *Pinicola enucleator* (Linnaeus)

The Pine Grosbeak breeds in evergreen forests of western states and across Canada. In New Jersey it is a rare winter visitant, occurring with infrequent winter invasion years. It is generally limited to northern counties, but there are individual records from southern New Jersey (e.g. Tuckerton and Cape May Point).

Invasions of varying strengths were noted in the following winters: 1836–1837, 1884–1885, 1899–1900, 1903–1904, 1929–1930, 1951–1952, 1961–1962, 1968–1969, 1976–1977, and 1980–1981.

There is a remarkable summer record of a female or an immature male in the Pine Barrens* on July 16, 1970.

A high count of 50 was made at Skyline Manor* on January 14, 1977.

PURPLE FINCH *Carpodacus purpureus* (Gmelin)

The Purple Finch breeds in coniferous forests of Pacific Coast states, southern Canada, and northeastern states. In New Jersey it is a migrant and winter visitant (October to early May), most evident during the fall. It is uncommonly a local breeder in northern counties (e.g. 2 broods at Stokes State Forest* in 1960). This species will adapt to nesting in a wide variety of ornamental evergreens.

Numbers of Purple Finches in the state vary greatly from year to year; the winter of 1972–1973 was a recent large invasion winter. A high count of 345 was obtained on the Walnut Valley Christmas count on December 15, 1979. A total of 1,096 was banded at Boonton during the spring of 1939. Perhaps the greatest invasion flight in recent decades was that of the fall of 1968, with remarkable flocks throughout.

HOUSE FINCH *Carpodacus mexicanus* (Müller)

The House Finch is a native of the United States and Mexico. It is now a permanent resident in New Jersey, apparently most numerous in winter. This western species has had a continual increase in abundance since its introduction into the metropolitan New York area in 1941. The first New Jersey record came from Ridgewood in Bergen County in 1949. A male banded at Union in the fall of 1958 was an attraction to almost 200 birders! By 1960 it seemed likely that the House Finch would eventually become a regular winter resident, and, indeed, it has become very common at feeding stations since its first appearance on the New Jersey Christmas counts in 1966. It may even reach pest proportions, and it should be remembered that this species is a serious agricultural problem in the West (e.g. feeding on buds of fruit trees).

The House Finch nests in a wide variety of habitats. In the East it often uses conifer trees in suburban landscapes. The clutch will usually be completed in late April or early May.

A high count of 1,177 was obtained on the Princeton Christmas count, December 28, 1980.

RED CROSSBILL *Loxia curvirostra* Linnaeus

The Red Crossbill breeds in pine forests of the western United States and Canada, southeast into New England. After invasion years it sometimes lingers to breed far south of its normal range. In New Jersey it is generally a rare winter visitor (December through March), but sightings have occurred in all months (especially after major invasion years). Nesting has been confirmed at a few locations (e.g. Ocean County in June 1941, and the Pine Barrens* of Burlington County, on July 24, 1956). Breeding has often been suspected at other locations (e.g. male feeding an immature at a Blairstown feeder in June 1970). Summer records from recently burned areas in the Pine Barrens* might also suggest nesting, but breeding at other seasons, especially spring, is also possible with this unpredictable species. The male will frequently feed the incubating female; his trips can help an observer find the nest.

Invasion years have included: winter 1961–1962, December 1963–February 1964, January 1966, fall 1968, fall 1969–March 1970, January 1972, January 1977, and winter 1981–1982.

WHITE-WINGED CROSSBILL *Loxia leucoptera* (Gmelin)

The White-winged Crossbill breeds in boreal forests of Alaska and Canada, south into New England. In New Jersey it is generally a rare winter visitant, even during invasion years. There is one probable breeding record at a Pine Barrens* lake in Ocean County where an adult was feeding fledged young in the summer of 1936. (The first attempted breeding of this species in New York State was reported in 1975.)

Invasion years are generally those listed for the Red Crossbill. In addition, December 1973 through February 1974 brought many reports. (The previous notable major flight was in the winter of 1952–1953.) The winter of 1981–1982 brought a huge invasion (e.g. several hundred were frequently observed near Newfoundland in Passaic County).

High counts include 100 at Oradell (March 9, 1965), 200 at West Creek (December 29, 1963), and 600 at Clinton Reservoir* (late January and early February 1982).

COMMON REDPOLL *Carduelis flammea* (Linnaeus)

The Common Redpoll breeds in tundra scrub of Alaska and northern Canada. In winter irregular movements bring some into the United States. In our area it is a winter visitant (December through April), usually rare to uncommon except in invasion years. This arctic species is generally restricted to northern parts of New Jersey, but there are infrequent sightings from southern counties as well.

Invasion winters have included: 1836–1837, 1878–1879, March 1888, 1899–1900, 1906–1907, 1908–1909, 1946–1947, February 1958, 1959–1960, February 1966, 1968–1969, 1969–1970, January 1972, February 1974, and January 1977.

Flocks of 2,000 or more were noted in the invasion of 1952–1953. Other high counts include 135 on a northern New Jersey Christmas count, December 26, 1971, and various counts of 100 or more (e.g. Pahaquarry, Franklin, West Orange).

HOARY REDPOLL *Carduelis hornemanni* (Holböll)

The Hoary Redpoll breeds in the high arctic of Alaska and Canada. In our region it is a very rare winter visitor, with invasions of the Common Redpolls. Field identification of this species can be difficult because there can be light-plumaged individuals of the Common Redpoll. There is, however, specimen and photographic documentation of the Hoary Redpoll in New Jersey. Recently there was 1 at Franklin Lakes, February 16–18, 1974, and there were about 20 statewide reports in the invasion year of 1977.

PINE SISKIN *Carduelis pinus* (Wilson)

The Pine Siskin breeds in much of the western United States, across Canada, and south through New England. In winter it has some southward movement, rarely to southeastern states.

In New Jersey the Pine Siskin is usually an uncommon winter visitor (October–April). Numbers are variable from year to year, and with some invasions it can be common, particularly in northern New Jersey.

There are a few breeding records for the state, primarily in our northern counties. In June and July 1981 immatures were noted in southern New Jersey, near Braddock's Mill Lake* in the Pine

Barrens.* Nests are found in evergreens, even in ornamental lawn trees (e.g. West Orange, summer of 1972). The breeding dates are variable but within the period of mid-March to May.

A pronounced invasion occurred in 1977 and early 1982. A high banding total of 467 was obtained at Cedar Run in the winter of 1969–1970.

AMERICAN GOLDFINCH *Carduelis tristis* (Linnaeus)

The American Goldfinch breeds in most of the northern United States, into Canada. In winter northern populations retreat, and goldfinches can be found south through Florida and Texas. In New Jersey the state bird is common as a permanent resident, nesting in open country throughout. Winter numbers are variable from year to year; in winter it particularly frequents thistle-seed feeders. It is one of our latest nesting species (August through September). The trim cup nest is usually well lined with thistledown, and while the female does the incubation she is fed by her mate.

EUROPEAN GOLDFINCH *Carduelis carduelis* (Linnaeus)

This species was introduced from Europe to the New York–New Jersey metropolitan areas in the nineteenth century, and there have been numerous subsequent releases and escapes. Some of the birds did become feral and established in several northeastern counties of our state, but records are few after 1915. Today there are still sightings (e.g. Basking Ridge in February 1981 at a feeder), but these are assumed to be recent cage-bird escapes.

A high count of 8 was made at Englewood on January 28, 1912.

EVENING GROSBEAK *Coccothraustes vespertinus* (Cooper)

The Evening Grosbeak breeds in conifers of western states and Canada, southeast into New England. Formerly this western species was only rarely reported in our region as an accidental winter visitor. Since 1950 it has expanded its range eastward, and it is now breeding regularly in New York State. In New Jersey it is now a migrant and winter visitor (October to May). It is seen most frequently in northern New Jersey, and in invasion years it can be quite common. It has bred twice in our state (in Morris and Sussex counties; both records in 1962).

The New Jersey history of this species began in March 1890 when a flock of 8 was recorded at Summit. In the winters of 1910–1911 and 1916–1917 there were invasions of several northern towns, and then 7 flight years into New Jersey were recorded in the next 16 winters. Recent peak invasions included the fall of 1959, December 1963 into February 1964, January 1966, and the fall of 1968.

High counts include 696 on December 1963 Christmas counts in northwestern New Jersey, and 726 on the Walnut Valley Christmas count in the winter of 1974–1975.

WEAVER FINCHES / *Passeridae*

HOUSE or ENGLISH SPARROW

Passer domesticus (Linnaeus)

The House Sparrow is a resident throughout the United States and southern Canada. In New Jersey it is a common permanent resident, at and near areas of human habitation. Apparently it was more numerous at the turn of the century when it was associated with an abundance of barns and horses. It has also had a recent population decline, possibly associated with the recent increases of the House Finch (an apparent competitor). The bulky grass nest is built by both sexes and is located in a wide variety of cavities or protected niches (often on buildings). Two or even three broods may be attempted between April and August.

The House Sparrow was introduced from Europe to North America in 1850 in New York City. It first appeared in New Jersey at Chatham in 1868. It has been widely introduced in many metropolitan areas throughout the world during the last hundred years.

Unestablished Exotics, Escapes, and Unsuccessful Introductions

Metropolitan areas frequently have a large pet trade, with concomitant escapes of exotic birds from both individual and zoological collections. When a zoo is involved escapes are carefully recorded and field reports of an exotic may often be traced back to its origin (e.g. a Yellow-billed Spoonbill seen at Cape May in 1981 was almost certainly a lost bird from a collection in Connecticut). However, when individuals lose or illegally release an exotic bird there is usually no way to trace field reports (e.g. many parrots), but the distant origins leave no doubt of their original captivity. Foreign waterfowl present particular problems since not only are they frequently kept in captivity by collectors, but individuals from some natural populations can wander the globe (e.g. Tufted Duck, Baikal Teal, Barnacle Goose, etc.). How, then, do we treat a field report of a European Wigeon or a West Indian Duck in New Jersey? Often its tameness may suggest a captive origin. And often there is simply no way to be sure if a wild bird is involved. This problem of natural versus escape status has been discussed in *New Jersey Audubon* (1980 6[2]:18–20), and in other recent articles in birding magazines.

This chapter provides a taxonomic list of some of the non-naturalized foreign birds that have been recorded in our state in recent years. Not included are introduced species that have become established as part of our permanent avifauna (e.g. Mute Swan, Ring-necked Pheasant, Rock Dove, Starling, and House Sparrow), as all such species are treated in the species account of the preceding chapter. For most species in the following list I have indicated the location and date of a report although there are often many reports.

Some of the following exotics may eventually become part of our avifauna, but most are ephemeral.

PENGUINS / *Spheniscidae*

MACARONI PENGUIN *Eudyptes chrysolophus*

It is native to the subantarctic islands in the Atlantic Ocean. One was caught swimming in Barnegat Bay* in the summer of 1981. It could not possibly have been a wild bird!

IBISES AND SPOONBILLS *Threskiornithidae*

SCARLET IBIS *Eudocimus ruber*

It is native to South America. This unmistakable bird was at Belvidere on August 31, 1965.

YELLOW-BILLED SPOONBILL *Platalea flavipes*

It is native to Australia. It was recorded flying over Cape May Point in September 1981 (see comments in introduction to this chapter).

DUCKS, GEESE, AND SWANS / *Anatidae*

BLACK SWAN *Cygnus atratus*

It is native to Australia. This species has been successfully introduced in eastern Quebec, and New Jersey birds raised young at Heislerville in the spring of 1976.

LESSER WHITE-FRONTED GOOSE *Anser erythropus*

It is native to Europe. See the species account; wild birds are possible.

BARNACLE GOOSE *Branta leucopsis*

It is native to Europe. See the species account; some records probably involved wild birds.

BAR-HEADED GOOSE *Anser indica*

It is native to India and Asia. Recorded at Point View Reservoir* in 1971 as an escape. (Other records are from Brigantine National Wildlife Refuge,* Johnson Park,* New Brunswick, etc.)

WHITE-FACED WHISTLING-DUCK *Dendrocygna viduata*

It is native to the West Indies. One was taken in Bergen County, October 1912.

MUSCOVY DUCK *Cairina moschata*

It is native to tropical America. It is a common domestic species and escapes are frequent. Forms include the natural dark plumage and all white varieties.

RUDDY SHELDUCK *Tadorna ferruginea*

It is native to Eurasia. Some of the nineteenth-century records may have involved wild birds, but recent ones are undoubtedly from waterfowl collections.

BAIKAL TEAL *Anas formosa*

It is native to Asia. This duck is an accidental visitor to West Coast states, but records from eastern states (e.g. Ohio, Pennsylvania, and New Jersey) are assumed to involve escapes. One was at Cape May Point in March 1961.

WHITE-CHEEKED PINTAIL *Anas bahamensis*

It is native to the Neotropics. There is a record from Hope Creek* in Salem County on November 3, 1980.

MANDARIN DUCK *Aix galericulata*

It is native to the Orient. A released bird was frequently seen at Silver Lake,* Belmar in the fall of 1972 and subsequently.

RED-CRESTED POCHARD *Netta rufina*

It is native to Eurasia. A bird was released at Silver Lake,* Belmar in the fall of 1972.

FALCONS / *Falconidae*

YELLOW-HEADED CARACARA *Milvago chimachima*

It is native to the Neotropics. One was seen over Summit in October 1967.

PHEASANTS / *Phasianidae*

CHUKAR *Alectoris chukar*

It is native to eastern Europe, but it has been established as an introduction in the western United States. Releases have been made in New Jersey on numerous occasions over many decades. Recently some were reported at Rossmoor in 1980.

GREEN PHEASANT *Phasianus versicolor*

It is native to Japan. This game bird has been released here at various locations in recent years.

CRANES / *Gruidae*

CROWNED CRANE *Balearica pavonina*

It is native to Africa. This spectacular bird has been common at zoological parks, and crane escapes are not infrequent. One was at Newport in September 1975.

GULLS AND TERNS / *Laridae*

ANDEAN GULL *Larus serranus*

It is native to South America. A bird is full breeding plumage was seen at the Great Swamp National Wildlife Refuge* by many observers in mid-April 1981.

PIGEONS AND DOVES / *Columbidae*

RINGED TURTLE-DOVE *Streptopelia risoria*

It is native to the Old World. This dove is commonly kept in captivity. It was seen at Hanover, December 1971.

PARROTS / *Psittacidae*

SULPHUR-CRESTED COCKATOO *Cacatua galerita*

It is native to Australasia. It was noted at Blairstown in the summer of 1971.

BUDGERIGAR *Melopsittacus undulatus*

It is native to Australia. This common pet is frequently seen as an escape (e.g. Kendall Park, September 15, 1974).

FISHER'S LOVEBIRD *Agapornis fisheri*

It is native to east Africa. One was present at Manalapan, October to November 1980.

ROSE-WINGED PARAKEET *Psittacula krameri*

It is native to India. There is a Somerdale report from January 1975.

BLOSSOM-HEADED PARAKEET *Psittacula roseata*

It is native to Asia. There is a Cape May record from September 1972.

ORANGE-FRONTED PARAKEET *Aratinga canicularis*

It is native to Central America. One was seen at Rutgers University, New Brunswick, in November 1974.

GOLDEN-CAPPED CONURE *Aratinga auricapilla*

It is native to Brazil. A Cape May sighting was made in the summer of 1972.

BLACK-HOODED PARAKEET *Nandayus nenday*

It is native to South America. There were various Mercer and Middlesex county reports in 1982. (This species is also called the Brown-hooded Conure.)

PATAGONIAN CONURE *Cyanoliseus patagonius*

It is native to South America. A report comes from Cape May in October 1979.

MONK PARAKEET *Myiopsitta monachus*

It is native to South America. See the species account; apparently few remain in New Jersey.

CANARY-WINGED PARAKEET *Brotogeris versicolurus*

It is native to South America. It was seen in New Brunswick in the summer of 1970.

CROWS AND JAYS / *Corvidae*

BLUE MAGPIE *Urocissa erythrorhyncha*

It is native to the Orient. It was seen at New Milford in December 1967.

TANAGERS / *Thraupinae*

BLUE-GRAY TANAGER *Thraupis virens*

It is native to Central and South America. One was seen repeatedly in Bergen County during 1973.

GROSBEAKS AND BUNTINGS *Cardinalinae*

RED-CAPPED CARDINAL *Paroaria gularis*

It is native to South America. One was at Cinnaminson during the winter of 1974–1975.

CROSSBILLS AND FINCHES
Fringillidae

EUROPEAN GOLDFINCH *Carduelis carduelis*

It is native to the Old World. See the species account for this temporarily successful cage-bird introduction.

EUROPEAN SISKIN *Carduelis spinus*

This Old World species is uncommon in captivity, but a probable escape was seen at a Bloomfield feeder in March 1983.

WEAVER FINCHES / *Passeridae*

ORANGE-CHEEKED WAXBILL *Estrilda melpoda*

It is native to Africa. It was seen as an escape in Teaneck in October 1967.

BLACK-RUMPED WAXBILL *Estrilda troglodytes*

It is native to Africa. This cage bird was seen with the preceding species at Teaneck.

Gazetteer
A Guide to Lesser Known Locations

Readers interested in the geographic location of place names mentioned in this volume may find the following gazetteer helpful. It lists all place names (other than townships) that are not generally indexed on standard road maps. These lesser known locations are listed in alphabetical order and when possible located by county and nearest city. All localities in the gazetteer are marked with an asterisk (*) whenever they occur in the text. Many of these locations are printed on road maps, but since they are often not indexed, they are much more easily located with the use of this gazetteer. It should be possible to find almost all other place names mentioned in the text on both map and index of any quality New Jersey road map. I used the official state transportation map (available from the New Jersey Division of Travel & Tourism, CN 384, Trenton, NJ 08625) for index checking. Standard abbreviations are used for compass directions (s.w. = southwest).

Absecon Bay, Absecon Inlet—Atlantic Co., s.e. of Absecon
Allamuchy Mountain State Park—Warren Co., e. of Allamuchy
Assunpink, Assunpink River, Assunpink Wildlife Management Area—Monmouth Co., e. of New Sharon
Avalon Causeway—Cape May Co., n. of Avalon
Baltimore Canyon—Atlantic Ocean, about 40 miles s.e. of Cape May Point
Barnegat Bay—Ocean Co., w. of Island Beach State Park
Barnegat Inlet—Ocean Co., n. of Barnegat Light
Beach Haven Inlet—Ocean Co., s. of Holgate
Bearfort Mountain, Bearfort Tower—Passaic Co., near West Milford
Bear Swamp—Cumberland Co., near Dividing Creek
Belleplain State Forest—Cape May Co., e. of Belleplain
Bidwell's Ditch—Cape May Co., near Reeds Beach
Big Flat Brook—Sussex Co., by Flatbrookville
Boonton Reservoir—Morris Co., s. of Boonton
Braddock's Mill Lake—Camden Co., near Lindenwold
Brigantine Beach, Brigantine Island—Atlantic Co., n. of Atlantic City
Brigantine National Wildlife Refuge, Brigantine Refuge—Atlantic Co., e. of Oceanville

Bull's Island State Park—Hunterdon Co., in the Delaware River, near Raven Rock
Cape May Canal—Cape May Co., s. of Cold Springs
Cape May Ferry—Cape May Co., from Cape May Point across the Delaware Bay to Lewes, Delaware
Cape May Inlet—Cape May Co., e. of Cape May
Cape May Lighthouse, Cape May Meadows—Cape May Co., near Cape May Point
Caven Cove—Hudson Co., e. of Jersey City
Cedar Run Lake—Ocean Co., s. of Manahawkin
Charles H. Rogers Refuge—Mercer Co., at Princeton
Clinton Reservoir, Clinton Road—Passaic Co., n. of Newfoundland
Colliers Mills Wildlife Management Area—Ocean Co., s.e. of Colliers Mills
Corson's Inlet—Cape May Co., n. of Strathmere
Culver's Gap—Sussex Co., near Normanock
Five Mile Beach—Cape May Co., at Wildwood Crest
Floodgates—Gloucester Co., on the Delaware River n. of Bridgeport
Great Bay—Atlantic Co., e. of Leeds Point
Great Channel—Cape May Co., n. of North Wildwood
Great Egg Harbor—Atlantic Co., s. of Somers Point
Great Piece Meadows—Morris Co., n.w. of Fairfield
Great Swamp, Great Swamp National Wildlife Refuge—Morris Co., s. of Green Village
Greenbrook Sanctuary—Bergen Co., e. of Tenafly
Hackensack Meadows—Bergen Co., e. of North Arlington
Hereford Inlet—Cape May Co., no. of North Wildwood
Higbee Beach—Cape May Co., n. of Cape May Point
High Point, High Point State Park—Sussex Co., w. of Colesville
Holgate Unit—Ocean Co., s. end of peninsula s. of Holgate
Hope Creek—Salem Co., w. of Hancocks Bridge
Hudson Canyon—Atlantic Ocean, about 80 mi. e. of Barnegat, at the edge of the continental shelf
Hutcheson Memorial Forest—Somerset Co., e. of East Millstone
Institute Woods—Mercer Co., at Princeton
Island Beach, Island Beach State Park—Ocean Co., peninsula s. of Seaside Park
Johnson Park—Middlesex Co., at New Brunswick
Kearny Marsh—Hudson Co., n.e. of Kearny
Kingsland—Bergen Co., s. of Rutherford
Kittatinny Mountains, Kittatinny Ridge—Sussex and Warren counties, mountains from High Point to the Delaware Water Gap
Kuser Bog—Sussex Co., in High Point State Park, w. of Colesville
Lake Assunpink—Monmouth Co., e. of New Sharon
Lake Girard—Sussex Co., near Beaver Lake
Lake Parsippany—Morris Co., e. of Parsippany

Lebanon State Forest—Burlington Co., n. of Chatsworth
Liberty State Park—Hudson Co., e. of Jersey City
Lily Lake—Cape May Co., Cape May Point
Mad Horse Creek Wildlife Management Area—Salem Co., s. of Salem
Mahlon Dickerson—Morris Co., near Petersburg
Manahawkin Wildlife Management Area—Ocean Co., e. of Manahawkin
Manasquan Inlet—Ocean Co., s. of Manasquan
Mannington Marsh—Salem Co., near Harrisonville
Metedeconk River—Ocean Co., s. of Metedeconk
Musconetcong River—Hunterdon Co., by Finesville
Newark Bay—Hudson Co., w. of Bayonne
Newark Meadows—Hudson Co., n. of Newark
New Jersey Audubon Scherman Sanctuary—Somerset Co., near Bernardsville
Nummy Island—Cape May Co., s. of Stone Harbor
Oradell Reservoir—Bergen Co., n.e. of Paramus
Overpeck Creek—Bergen Co., w. of Palisades Park
Pedricktown Marshes—Salem Co., near Pedricktown
Pequannock Watershed—Passaic Co., n. of Pompton Lakes
Petty's Island—Camden Co., in the Delaware River s. of Camden
Pine Barrens—Centered in Burlington Co. and Atlantic Co.; a regional ecosystem of south-central New Jersey, dominated by pitch pines and oaks
Pohatcong Creek Valley—Warren Co., e. of Phillipsburg
Point View Reservoir—Passaic Co., near Wayne
Raccoon Ridge—Warren Co., near Pahaquarry
Ramapo Mountains—Bergen Co., n. of Ramsey
Raritan Arsenal—Middlesex Co., s. of Nixon
Raritan Bay—waters between Staten Island, New York, and New Jersey, w. of Sandy Hook
Raritan Estuary, Raritan River—Middlesex Co., n. of Sayreville
Rifle Camp Park—Passaic Co., n. of Paterson
Rosedale Park—Mercer Co., s.w. of Princeton
Round Valley Reservoir—Hunterdon Co., w. of White House
Scudder's Falls—Mercer Co., in Delaware River n. of Trenton
Secaucus Marshes—Hudson Co., near Secaucus
Seven Mile Beach—Cape May Co., e. of Stone Harbor
Shark River, Shark River Inlet—Monmouth Co., n. of Belmar
Silver Lake—Monmouth Co., at Belmar
Skyline Manor—Passaic Co., Ringwood State Park
Smithville Pond—Atlantic Co., at Smithville
South Mountain Reservation—Essex Co., n. of Millburn
Split Rock Reservoir—Morris Co., near Rockaway
Spruce Run Reservoir—Hunterdon Co., w. of Clinton
Stokes State Forest—Sussex Co., n.w. of Culvers Inlet
Sunrise Mountain—Sussex Co., n. of Culvers Inlet

Swartswood Lake—Sussex Co., s.w. of Swartswood
Sylvan Lake—Monmouth Co., at Avon-By-The-Sea
Trenton Marshes—Mercer Co., s. of Trenton
Troy Meadows—Morris Co., near Parsippany-Troy Hills
Turkey Swamp—Monmouth Co., w. of Georgia
Turtle Back Zoo—Essex Co., at West Orange
Wallkill River—Sussex Co., e. of Sussex
Watchung Reservation—Union Co., w. of Mountainside
Wawayanda State Park—Sussex Co., near Cherry Ridge
Week's Landing—Cape May Co., w. of Goshen
Wharton State Forest, Wharton Tract—Burlington Co., n. of Batsto
Yards Creek—Warren Co., n.w. of Walnut Valley

Bibliography
With Emphasis on Recent Literature

Adams, Bruce. 1970. Island Beach spring project—1969. *EBBA News* 33:32–34.

Adams, Bruce, and John Miller. 1975. The Absecon Bay heron colony. *EBBA News* 38:103–108.

Akers, James F. 1981. *All year birding in southern New Jersey*. Pomona: Stockton State College.

Arbib, R. S., O. S. Pettingill, and S. H. Spofford. 1966. *Enjoying birds around New York City*. Boston: Houghton Mifflin Co.

Bacinski, Peter. 1983a. North central Jersey Highlands—breeding bird survey 1980. *Urner Field Observer* 18:6–16.

———. 1983b. North central Jersey Highland breeding bird survey 1981. *Urner Field Observer* 18:17–26.

Baird, Donald. 1967. Age of fossil birds from the greensands of New Jersey. *Auk* 84:260–262.

Baird, James. 1962. A prolonged Starling flight. *Wilson Bulletin* 74:187–189.

Baird, John W. 1980. The selection and use of fruit by birds in an eastern forest. *Wilson Bulletin* 92:63–73.

Basakowski, Thomas. 1982. Status and habitat distribution of raptors in the Great Swamp Wilderness Area. *New Jersey Audubon* 8(4):63–67.

Berzinger, John, and Robert Speiser. 1982. Breeding birds of New Jersey's Ramapo Mountains. *New Jersey Audubon* 8(2):24–26.

Bihun, Andrew, Jr. 1972. Montclair Bird Club hawk watch. *New Jersey Nature News* 27:115–118.

Black, Irving H. 1970. Past and present status of the birds of the lower Hackensack River marshes. *New Jersey Nature News* 25:57–70.

Black, Irving H., and William B. Marx. 1971. Northern New Jersey Highland breeding bird survey. *Urner Field Observer* 13:19–26, and 14:22–28.

Bongiorno, S. F., and Jeff Swinebroad. 1969. Increase in Herring Gull colony in Cape May, New Jersey. *Wilson Bulletin* 81:99–100.

Bordner, D. L. 1964. Island Beach Operation Recovery 1963. *EBBA News* 27:101–106.

Boyle, William J., Jr. 1979. *New Jersey field trip guide*. Summit: Summit Nature Club.

Brady, Alan. 1976. A summary of recent pelagic trips off the New Jersey coast. *Cassinia* 56:7–14.

Bull, John. 1964. *Birds of the New York area.* New York: Harper & Row.
———. 1974. *Birds of New York State.* New York: Doubleday. Supplement. 1976. Federation of New York State Bird Clubs.
Burger, Joanna. 1976. Daily and seasonal activity patterns in breeding Laughing Gulls. *Auk* 93:308–323.
———. 1977. Nesting behavior of Herring Gulls: Invasion into *Spartina* salt marsh areas of New Jersey. *Condor* 79:162–169.
———. 1978a. Competition between Cattle Egrets and native North American herons, egrets, and ibises. *Condor* 80:15–23.
———. 1978b. Great Black-backed Gulls breeding in salt marsh in New Jersey. *Wilson Bulletin* 90:304–305.
———. 1979a. Colony size: A test for breeding synchrony in Herring Gull *(Larus argentatus)* colonies. *Auk* 96:694–703.
———. 1979b. Competition and predation: Herring Gulls versus Laughing Gulls. *Condor* 81:269–277.
———. 1980. Territory size differences in relation to reproductive stage and type of intruder in Herring Gulls *(Larus argentatus). Auk* 97:733–741.
———. 1981. Feeding competition between Laughing Gulls and Herring Gulls at a sanitary landfill. *Condor* 83:328–335.
Burger, J., M. A. Howe, D. C. Hahn, and J. Chase. 1977. Spatial and temporal patterns of foraging in migrating shorebirds in a tidal regime. *Auk* 94:743–758.
Burger, Joanna, and Fred Lesser. 1979. Breeding behavior and success in salt marsh Common Tern colonies. *Bird-Banding* 50:322–337.
———. 1980. Nest site selection in an expanding population of Herring Gulls. *Bird-Banding* 51:270–280.
Burger, Joanna, and Joseph Shishler. 1978. Nest-site selection of Willets in a New Jersey salt marsh. *Wilson Bulletin.* 90:599–607.
———. 1979. Immediate effects of ditching a marsh on breeding Herring Gulls. *Biological Conservation* 15:85–104.
———. 1980. Colony and nest site selection in Laughing Gulls in response to tidal flooding. *Condor* 82:251–258.
Burk, Dick. 1980. Summit club counts. *New Jersey Audubon* 6(1):9–13.
Caccamise, Donald F. 1977. Breeding success and nest site characteristics of the Red-winged Blackbird. *Wilson Bulletin* 89:396–403.
———. 1978. Seasonal patterns of nesting mortality in the Red-winged Blackbird. *Condor* 80:290–294.
Choate, Ernest A. 1972. Spectacular hawk flight at Cape May Point, New Jersey on 16 October 1970. *Wilson Bulletin* 84:340–341.
Choate, Ernest A., and J. d'Arcy Northwood. 1969. Forty years of Christmas bird counts in Cape May County (1927–1966). *Cassinia* 51:3–10.
Clark, William S. 1972. Migration trapping of hawks (and owls) at Cape May, N.J.—fifth year. *EBBA News* 35:121–131.
———. 1974. Second record of the Kestrel *(Falco tinnunculus)* for North America. *Auk* 91:172.

———. 1976. Cape May Point raptor banding station—1974 results. *North American Bird Bander* 1:5–13.

———. 1978. Spring hawk movement at Sandy Hook, N.J.—1977. *New Jersey Audubon* 4(2):43–47.

Dennis, John V. 1981. A summary of banded North American birds encountered in Europe. *North American Bird Bander* 6:88–96.

Dowdell, James. 1980. The 1980 spring hawk watch at Sandy Hook, New Jersey. *Peregrine Observer* 3(2):8–10.

Dunne, Peter J. 1977a. Spring hawk movement along Raccoon Ridge, N.J.—1976. *New Jersey Audubon* 3(2):20–29.

———. 1977b. Summary of the 1977 hawk flight at Cape May Point, N.J. *Cassinia* 57:21.

Dunne, Peter J., and William S. Clark. 1977. Fall hawk movements at Cape May Point, N.J.—1976. *New Jersey Audubon* 3(7–8):114–124.

Dunne, Peter, David Sibley, Clay Sutton, and Wade Wander. 1982. 1982 Aerial shorebird survey of Delaware Bay. *New Jersey Audubon* 8(4):68–75.

Fables, David G., Jr. 1955. *Annotated list of New Jersey birds*. Newark: Urner Ornithological Club.

———. 1962. Breeding birds of the New Jersey Pine Barrens. *New Jersey Nature News* 17:60–64.

Foy, Dorothy [Mrs. Roger W.]. 1976. Recoveries and foreign retraps of birds banded at Island Beach, N.J. *Bird-Banding* 47:214–230.

———. 1981. Yellow-headed Blackbird bandings in N.J. *North American Bird Bander* 6:16–17.

Galli, Anne E., Charles F. Leck, and Richard T. T. Forman. 1976. Avian distribution patterns in forest islands of different sizes in central New Jersey. *Auk* 93:356–364.

Galli, Joan, and Richard Kane. 1981. 1979 colonial waterbird populations of New Jersey. *New Jersey Audubon* 7(3):36–43.

Greenlaw, Jon S. 1978. The relation of breeding schedule and clutch size to food supply in the Rufous-sided Towhee. *Condor* 80:24–33.

Harding, John J., and Justin J. Harding. 1980. *Birding the Delaware Valley region*. Philadelphia: Temple University Press.

Harrison, Hal H. 1975. *A field guide to birds' nests*. Boston: Houghton Mifflin Co.

Heintzelman, Donald S. 1970. *Hawks of New Jersey*. Trenton: New Jersey State Museum Bulletin 13.

———. 1971a. Age, sex, and wing lengths of autumn migrants from central New Jersey. *Science Notes*, no. 1. Trenton: New Jersey State Museum.

———. 1971b. Birds of the New Land Research Reserve. *Science Notes*, no. 5. Trenton: New Jersey State Museum.

———. 1972a. Age classes of birds migrating southward through central New Jersey—autumn 1970. *EBBA News* 35:239–244.

———. 1972b. Banding recoveries from the records of Elise and Stanley Dickerson. *EBBA News* 35:3–28.

———. 1972c. *A guide to northeastern hawk watching*. Published privately.
Kane, Richard. 1976. The 1975 fall shorebird migration in the Hackensack Meadowlands. *New Jersey Audubon* 2(2):8–14.
———. 1978. Birds of the Kearny Marsh. *New Jersey Audubon* 4(5):22–27.
Kane, Richard, and Richard B. Farrar. 1976. 1976 coastal colonial bird survey of New Jersey. *New Jersey Audubon* 2(11):8–14.
Kane, Richard, and William B. Marx. 1972. Northern New Jersey Highland breeding bird survey 1971–72. *Urner Field Observer* 14:5–21.
Keith, Allan R. 1983. History of the publications of the Urner Ornithological Club. *Urner Field Observer* 18:27–34.
Kennard, John H. 1975. Longevity records of North American birds. *Bird-Banding* 46:55–73.
Koebel, Timothy D. 1983a. New Jersey inland—the fall shorebird flight. *Urner Field Observer* 18:40–48.
———. 1983b. The origin and history of Raccoon Ridge. *Urner Field Observer* 18:49–52.
Kunkle, Donald E. 1976. Some bird observations on, over, and around Delaware Bay. *Urner Field Observer* 16:13–25.
Leck, Charles F. 1972a. The Monk Parakeet in New Jersey. *Cassinia* 53:50.
———. 1972b. Wave phenomena of land migrants at Island Beach State Park, New Jersey. *Bird-Banding* 43:20–25.
———. 1973a. Attempts of gulls to rob a Common Egret and an Osprey. *Cassinia* 54:32.
———. 1973b. Erythrism and melanism in the Barn Owl. *Cassinia* 54:32.
———. 1975a. A bird-banding report from Hutcheson Memorial Forest: A summary of three seasons. *EBBA News* 38:131–132.
———. 1975b. *The birds of New Jersey: Their habits and habitats*. New Brunswick: Rutgers University Press.
———. 1975c. Changes in the birdlife of Princeton, New Jersey: A comparison of 1900 and 1974. *Cassinia* 55:29–32.
———. 1977. 1970–76 New Jersey Christmas Counts. *New Jersey Audubon* 3(2):15.
———. 1979a. A comparison of two winters at a suburban feeding station. *North American Bird Bander* 4:34–35.
———. 1979b. Subspecies for New Jersey observers. *New Jersey Audubon* 5(3):37–39.
———. 1982. Declines in some forest birds over twenty years. *New Jersey Audubon* 8(1):5.
Leck, Charles F., Sharon A. Brady, and Wade Wander. 1982. The effects of canopy opening on breeding and migrant birds at Hutcheson Memorial Forest. *William L. Hutcheson Memorial Forest Bulletin* 6(2):20–23.
Leck, Charles F., and Phyllis Hawkins. 1977. Observations of shorebirds at selected New Jersey sewage lagoons. *Cassinia* 57:3–5.
Leck, Charles F., Bertram G. Murray, Jr., and Jeff Swinebroad. 1981. Changes in breeding bird populations at Hutcheson Memorial Forest since 1958. *William L. Hutcheson Memorial Forest Bulletin* 6:8–15.

Marx, William B. 1971a. A history of the Barnegat Christmas Count (1925–1970). *Urner Field Observer* 13:35–52.

———. 1971b. The 1970 spring migration of hawks in the central highlands of New Jersey. *Urner Field Observer* 13:11–17.

Miller, John C. 1977. Data on bird banding at Island Beach State Park, New Jersey. *Cassinia* 57:10–18.

Murray, Bertram G., Jr. 1971. A small Great Crested Flycatcher: A problem in identification. *Bird-Banding* 42:119.

———. 1972. Some rare specimens of New Jersey birds. *Cassinia* 53:27–28.

———. 1976. The return to the mainland of some nocturnal passerine migrants over the sea. *Bird-Banding* 47:345–358.

———. 1979. Fall migration of Blackpoll and Yellow-rumped Warblers at Island Beach, N.J. *Bird-Banding* 50:1–11.

Perrone, Steve, ed. 1982. *Guide to wildlife management areas*. Trenton: Department of Environmental Protection.

Peterson, Roger T. 1980. *A field guide to the birds*. Boston: Houghton Mifflin Co.

Pettingill, Olin S. 1977. *A guide to bird finding east of the Mississippi*. New York: Oxford University Press.

Poole, Earl L. 1964. *Pennsylvania birds*. Narberth, Pa.: Livingston Publishing Co.

Prescott, Kenneth W. 1972a. Screech Owl kills netted sparrows. *Bird-Banding* 43:143.

———. 1972b. Three-year banding project of Tree Swallows. *EBBA News* 35:244–249.

Price, Katherine G. 1971. Island Beach bird banding station. *EBBA News* 34:250–252.

Rogers, Charles H. 1961. Increase in numbers in New Jersey's coastal birds in fifty years. *New Jersey Nature News* 16:124–134.

Ryan, Richard. 1980. Exotic waterfowl in New Jersey. *New Jersey Audubon* 6(2):18–20.

Schwab, Don. 1981. Fall 1979 banding at Great Swamp. *New Jersey Audubon* 7(1):2–6.

Shields, William M., Thomas C. Grubb, Jr., and Anthony Telis. 1974. Use of native plants by Monk Parakeets in New Jersey. *Wilson Bulletin* 86:172–173.

Soucy, Leonard J., Jr. 1976. Raptor report—Kittatinny Mountains. *North American Bird Bander* 1:108–113.

———. 1980a. *New Jersey's owls*. Published privately.

———. 1980b. Three long-distance recoveries of banded N.J. Barn Owls. *North American Bird Bander* 5:97.

———. 1983. First Gyrfalcon banded in New Jersey. *New Jersey Audubon* 9(1):2.

Stone, Witmer. 1909. *The birds of New Jersey*. Trenton: Annual Report of the New Jersey State Museum 1908.

———. [1937] 1965. *Bird studies of Old Cape May*. Reprint. New York: Dover.

Sussman, Oscar et al. 1971. Nine years of bird banding at four areas in New Jersey. *EBBA News* 34:85–88.

Sutton, Clay C., Jr. 1981. The breeding birds of the Bear Swamp. *Peregrine Observer* 4(2), unpaginated.

Swinebroad, Jeff. 1962. An annotated list of the birds of Hutcheson Forest. *Bulletin of the New Jersey Academy of Science* 7:1–6.

———. 1963. Edge effect and bird populations in southern New Jersey. *New Jersey Nature News* 18:66–73.

Swinebroad, Jeff, and Oscar Sussman. 1965. Breeding bird populations in New Jersey. *New Jersey Nature News* 20:102–111.

Tilly, Fred. 1973. The 1972 autumn hawk count at Raccoon Ridge, Warren County, New Jersey. *Science Notes,* no. 13. Trenton: New Jersey State Museum.

———. 1979. The 1978 Raccoon Ridge autumn hawkwatch. *New Jersey Audubon* 5(1):1–2.

Tischner, Kenneth. 1977. Road kills of south Jersey. *Cassinia* 57:22–23.

Wander, Wade. 1977. Breeding birds of Sandy Hook—1976. *New Jersey Audubon* 3(5–6):84–90.

———. 1979. Waterbird concentrations at the Assunpink Fish and Wildlife Area, 1974–78. *New Jersey Audubon* 5(2):17–24.

———. 1980. Breeding birds of southern New Jersey cedar swamps. *New Jersey Audubon* 6(4):51–65.

———. 1982. Breeding status of grassland birds in New Jersey. *New Jersey Audubon* 8(1):2–4.

Wander, Wade, and Peter Dunne. 1982. Species and numbers of shorebirds on the Delaware bayshore of New Jersey—spring 1981. *New Jersey Audubon* 7(4):59–64.

Ward, David, Jr. 1980a. Autumn 1979 seabird watch in Avalon. *New Jersey Audubon* 6(1):2–4.

———. 1980b. Operation seawatch, fall 1979. *Peregrine Observer* 3(1), unpaginated.

Welton, Melinda. 1983. The status of Cliff Swallows in New Jersey. *New Jersey Audubon* 9(1):4–5.

Willard, David E. 1977. The feeding ecology and behavior of five species of herons in southeastern New Jersey. *Condor* 79:462–470.

Wood, Merrill. 1979. *Birds of Pennsylvania.* University Park: Pennsylvania State University Press.

Index